大数据创新人才培养系列

Development Practice on Redis 6

Redis 6 开发与实战

张云河 王硕 / 编著

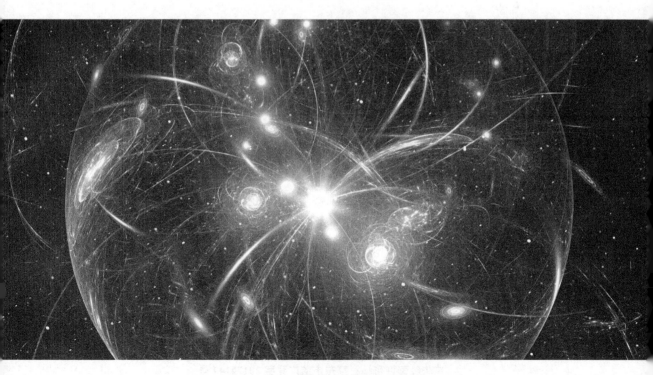

人民邮电出版社

北京

图书在版编目（CIP）数据

Redis 6 开发与实战 / 张云河，王硕编著. -- 北京：
人民邮电出版社，2021.8
（大数据创新人才培养系列）
ISBN 978-7-115-56321-7

Ⅰ．①R… Ⅱ．①张… ②王… Ⅲ．①数据库—基本知识 Ⅳ．①TP311.138

中国版本图书馆CIP数据核字(2021)第063083号

内 容 提 要

  Redis 是一个十分热门的内存数据库，它拥有众多优秀特性，已经被越来越多的公司采用，值得每一位开发者学习。本书主要包括：初识 Redis、Redis 常用数据类型及命令、Redis 常用命令、Redis 高级主题、Redis 缓存的持久化、Redis 集群环境部署、Redis 开发实战、Spring Boot 与 Redis 整合应用、Redis 监控、Redis 的缓存设计与优化、扩展知识等内容。通过本书讲解 Redis 在开发和运维等方面的实例，读者不仅可以系统地学到有关 Redis 的大量知识，还可以将实例中所讲解的内容用于开发和运维等各类生产实践。本书实例涉及的各种知识、命令和工具，均来自作者对一线企业级应用的总结。本书的特色在于讲解知识点的同时，辅以大量生动的实例，以帮助读者更好地理解 Redis。

  本书通过理论和实例全面系统地讲解 Redis，适合所有渴望深入学习 Redis 的读者阅读。

◆ 编 著 张云河 王 硕
  责任编辑 刘 博
  责任印制 王 郁 马振武

◆ 人民邮电出版社出版发行 北京市丰台区成寿寺路11号
 邮编 100164 电子邮件 315@ptpress.com.cn
 网址 https://www.ptpress.com.cn
 固安县铭成印刷有限公司印刷

◆ 开本：787×1092 1/16
 印张：17.5   2021年8月第1版
 字数：424千字   2024年8月河北第5次印刷

定价：59.80元

读者服务热线：(010)81055256 印装质量热线：(010)81055316
反盗版热线：(010)81055315
广告经营许可证：京东市监广登字 20170147 号

前言

党的二十大报告中提到："教育、科技、人才是全面建设社会主义现代化国家的基础性、战略性支撑。"在教育改革、科技变革等背景下，信息技术领域的教学发生着翻天覆地的变化。

为什么要写这本书

Redis 作为一个流行的 key-value 内存数据库，由于性能高、数据类型丰富、API 功能强大、可用性高等特点，已经被越来越多的企业用于生产实践。Redis 可以将所有数据都存放在内存中，所以它的读写性能非常惊人；Redis 还可以将内存中的数据利用快照和日志的形式保存到硬盘上，保证内存中的数据不会丢失，等等。总之，在合适的场景使用好 Redis，它就会像一把瑞士军刀一样方便。

本书基于 Redis 6。为帮助读者理解，书中使用了大量的实例，几乎涵盖了 Redis 的方方面面，从 Redis 基本数据类型、常用命令等基础知识，再到 Redis 缓存持久化、集群环境部署和开发实战等高级主题。

学习任何技术都要理论联系实践，所以本书将通过大量的实例向读者系统地讲解 Redis 的各个知识点。在读者的个人计算机上，只要遵循本书实例的操作步骤，都可以很容易地理解每个实例的知识点，缩短学习 Redis 的时间和降低学习编程的难度。

希望本书能让读者更好地理解 Redis，并能帮助读者在工作中正确使用 Redis 的新特性。

本书有何特色

1．涵盖 Redis 核心知识点

本书基于 Redis 6 进行讲解，涵盖了 Redis 的环境搭建、常用数据类型及命令、常用命令、服务器配置、事务、缓存持久化，以及 Redis 6 的集群环境搭建和编程开发实战等，可以帮助读者全面掌握 Redis 的知识点。

2．实战实例丰富

本书涵盖了 Redis 涉及的各个方面，对 Redis 的日常使用方法进行了全面讲解和技巧提

Redis 6 开发与实战

示，使用的实例来源于笔者在企业开发和运维工作中的经验总结。

3．使用 Java 操作 Redis

本书使用 Java 操作 Redis，并使用 Spring Boot 2 框架与 Redis 进行集成。书中对环境的搭建和代码操作的步骤都进行了详细的解释。

4．提供完善的技术支持和售后服务

由于笔者水平有限，书中难免存在疏漏和不足之处，敬请读者批评指正。如有任何问题欢迎发送邮件至邮箱：xpws2006@163.com。读者在阅读本书的过程中有任何疑问，都可以通过该邮箱获得帮助，也可以加入本书的 QQ 群（629588406）：Redis 高级开发群。

本书内容及知识体系

第 1 章　初识 Redis

本章主要介绍了 Redis 服务器的安装和基本操作，包括在 Windows 和 Linux 下启动和停止 Redis、使用 redis-cli 连接到 Redis 服务器等。在本章的最后使用 VMware 安装 6 个 CentOS 7 操作系统，搭建了 Redis 集群环境。

第 2 章　Redis 常用数据类型及命令

本章主要介绍了 Redis 的常用数据类型和操作数据类型的常用命令，还介绍了 Redis HyperLogLog 的功能。

第 3 章　Redis 常用命令

本章主要介绍了 Redis 的常用命令，包括键值相关命令和服务器相关命令。

第 4 章　Redis 高级主题

本章主要介绍了 Redis 的高级主题，包括 Redis 服务器配置、Redis 事务、Redis 发布和订阅、Pipeline 批量发送请求、数据备份与恢复、Redis 性能测试、Redis 客户端连接和 Redis 服务开机自启动。

第 5 章　Redis 缓存的持久化

本章主要介绍了 Redis 中的两种持久化方式 RDB 和 AOF，讲解了在 Redis 中如何启动 RDB 和 AOF 来实现持久化，并揭示了持久化的工作原理。此外，本章讨论了 RDB 和 AOF 之间的区别，以及如何将这两种持久化方式结合起来使用。

第 6 章　Redis 集群环境部署

本章介绍了 Redis 集群环境部署，包括主从复制、哨兵模式和 Redis 集群。其中主要介绍了 Redis 6 集群配置的详细实验步骤和工作原理。

第 7 章　Redis 开发实战

本章主要介绍了如何使用 Java 开发 Redis 程序，讨论了 Redis 的数据类型和 API 的使用方法，讲解了使用 Redis 客户端 Jedis 应用程序的实例，并使用实例介绍了 Redis 在生产环境中的使用场景和技巧。

本章在 Windows 下安装了 Java 和 Tomcat，搭建了 Redis 的 Java 开发环境，介绍了使用 Java 操作 Redis；在 Linux 下安装了 Java、Tomcat 和 Nginx，介绍了在 Tomcat 上使用 Redis 保存 Session。

第 8 章　Spring Boot 与 Redis 整合应用

本章介绍了 Spring Boot 2 框架与 Redis 的集成，详细介绍了 RedisTemplate API 的使用方法，并使用实例介绍了在 Spring Session 中使用 Redis 的场景和原理。

第 9 章　Redis 监控

本章主要介绍了 Redis 的性能监控和自定义开发应用程序来监控 Redis 的性能，其中使用的是 WebSocket 技术与后台进行消息推送和接收。

第 10 章　Redis 的缓存设计与优化

本章主要介绍了 Redis 缓存的优点和缺点，以及在生产环境下出现缓存雪崩与缓存穿透的原因和解决方法。本章还介绍了布隆过滤器。

第 11 章　扩展知识

本书涉及的技术内容比较多，所以把读者需要掌握的内容单独汇聚成一章，包括配置 CentOS 7、Maven 基础知识、配置 IntelliJ IDEA、使用 VMware、配置 SecureCRT、Chrome 的常用技巧。

本章介绍在 Windows 和 Linux 下安装 Python 3，并使用 Redis 模块操作 Redis 6 的单机和集群。

适合阅读本书的读者

- 需要全面学习 Redis 的人员。
- 对 Redis 编程感兴趣的人员。
- 希望提高 Redis 使用水平的程序员。
- 开设相关课程的高校师生。

感谢

感谢我的家人一直以来对我的支持和宽容。感谢我的妻子，在家庭方面付出了很多，做出了重要的贡献，让我安心写作。感谢我的两个女儿，你们的微笑是爸爸消除疲惫的良药，爱你们。

本书的顺利出版要感谢人民邮电出版社的所有编校人员，感谢他们在选题策划和稿件整理方面做出的大量工作，再次感谢他们对我的信任和支持。

最后祝福所有读者在职场一帆风顺，事业有成。

本书资源下载

本书的实例使用 Git 来管理程序代码和版本，为了方便大家获取实例代码，代码均托管在 GitHub 上，链接地址可在人邮教育社区（www.ryjiaoyu.com）搜索本书找到。如果实例代码在本书出版后还有更新，那么将会更新到相关链接对应的 GitHub 库上。

读者安装好 Git 环境后，可以使用 git clone https://github.com/cxinping/Redis 命令，把本书托管到 GitHub 上的实例代码下载到本地硬盘上，下载后的实例代码如下所示。

名称	修改日期	类型	大小
Chapter01	2020/11/29 0:09	文件夹	
Chapter04	2020/11/29 0:09	文件夹	
Chapter06	2020/11/29 0:09	文件夹	
Chapter07	2020/11/29 0:37	文件夹	
Chapter08	2020/11/29 0:09	文件夹	
Chapter09	2020/11/29 0:10	文件夹	
Chapter10	2020/11/29 0:10	文件夹	
Chapter11	2020/11/29 0:10	文件夹	
doc	2020/11/29 0:41	文件夹	
Tools	2020/11/29 0:41	文件夹	
README.md	2020/11/29 0:10	MD 文件	9 KB

笔者

2022 年 12 月

目录

第1章 初识Redis ·············· 1
1.1 Redis快速入门 ············ 1
1.1.1 Redis简介 ·············· 1
1.1.2 Redis特性 ·············· 2
1.2 Redis环境搭建 ············ 3
1.2.1 在Windows下安装Redis ····· 4
1.2.2 在Linux下安装Redis ······· 7
1.3 Redis可视化工具 ·········· 13
1.4 搭建Redis集群环境 ········ 14
1.4.1 配置VMware准备安装 CentOS ················ 15
1.4.2 安装Linux ·············· 18
1.4.3 安装VMware Tools ······· 20
1.4.4 虚拟机与宿主机的网络设置 ··· 21
1.4.5 复制虚拟机 ·············· 25

第2章 Redis常用数据类型及命令 ···· 28
2.1 String类型 ················ 28
2.1.1 SET ···················· 28
2.1.2 SETNX ················ 29
2.1.3 SETEX ················ 29
2.1.4 SETRANGE ············ 30
2.1.5 MSET ················ 30
2.1.6 MSETNX ·············· 31
2.1.7 APPEND ·············· 31
2.1.8 GET ·················· 32
2.1.9 MGET ················ 32
2.1.10 GETRANGE ·········· 33

2.1.11 GETSET ·············· 33
2.1.12 STRLEN ·············· 34
2.1.13 DECR ················ 34
2.1.14 DECRBY ·············· 35
2.1.15 INCR ················ 35
2.1.16 INCRBY ·············· 36
2.2 Hash类型 ················ 36
2.2.1 HSET ················ 36
2.2.2 HSETNX ·············· 37
2.2.3 HMSET ·············· 37
2.2.4 HGET ················ 37
2.2.5 HMGET ·············· 38
2.2.6 HGETALL ············ 38
2.2.7 HDEL ················ 38
2.2.8 HLEN ················ 39
2.2.9 HEXISTS ·············· 39
2.2.10 HINCRBY ············ 40
2.2.11 HKEYS ·············· 40
2.2.12 HVALS ·············· 41
2.3 List类型 ················ 41
2.3.1 LPUSH ················ 41
2.3.2 LPUSHX ·············· 42
2.3.3 RPUSH ················ 42
2.3.4 RPUSHX ·············· 43
2.3.5 LPOP ················ 43
2.3.6 RPOP ················ 44
2.3.7 LLEN ················ 44
2.3.8 LREM ················ 45

2.3.9	LSET	46
2.3.10	LTRIM	47
2.3.11	LINDEX	48
2.3.12	LINSERT	49
2.3.13	RPOPLPUSH	49

2.4 Set 类型 50
 2.4.1 SADD 51
 2.4.2 SREM 51
 2.4.3 SMEMBERS 52
 2.4.4 SCARD 52
 2.4.5 SMOVE 53
 2.4.6 SPOP 53
 2.4.7 SRANDMEMBER 54
 2.4.8 SINTER 55
 2.4.9 SINTERSTORE 55
 2.4.10 SUNION 56
 2.4.11 SUNIONSTORE 56
 2.4.12 SDIFF 57
 2.4.13 SDIFFSTORE 57

2.5 Sorted Set 类型 58
 2.5.1 ZADD 58
 2.5.2 ZREM 59
 2.5.3 ZCARD 60
 2.5.4 ZCOUNT 61
 2.5.5 ZSCORE 61
 2.5.6 ZINCRBY 62
 2.5.7 ZRANGE 62
 2.5.8 ZREVRANGE 63
 2.5.9 ZREVRANGEBYSCORE 64
 2.5.10 ZRANK 64
 2.5.11 ZREVRANK 65
 2.5.12 ZREMRANGEBYRANK 65
 2.5.13 ZREMRANGEBYSCORE 66
 2.5.14 ZINTERSTORE 67
 2.5.15 ZUNIONSTORE 68

2.6 Redis HyperLogLog 69
 2.6.1 Redis HyperLogLog 常用命令 69
 2.6.2 Redis HyperLogLog 实例 69

第 3 章 Redis 常用命令 71

3.1 键值相关命令 71
 3.1.1 KEYS 71
 3.1.2 SCAN 72
 3.1.3 EXISTS 73
 3.1.4 DEL 73
 3.1.5 EXPIRE 73
 3.1.6 TTL 74
 3.1.7 SELECT 74
 3.1.8 MOVE 74
 3.1.9 PERSIST 75
 3.1.10 RANDOMKEY 75
 3.1.11 RENAME 75
 3.1.12 TYPE 76

3.2 服务器相关命令 76
 3.2.1 PING 76
 3.2.2 ECHO 76
 3.2.3 QUIT 76
 3.2.4 DBSIZE 76
 3.2.5 INFO 77
 3.2.6 MONITOR 79
 3.2.7 CONFIG GET 80
 3.2.8 FLUSHDB 80
 3.2.9 FLUSHALL 80

第 4 章 Redis 高级主题 81

4.1 服务器配置 81
 4.1.1 Redis 服务器允许远程主机访问 81
 4.1.2 客户端远程连接 Redis 服务器 82
 4.1.3 设置密码 82
 4.1.4 Redis 端口修改 83
 4.1.5 查看配置 84
 4.1.6 修改配置 84
 4.1.7 配置项说明 84

4.2 Redis 事务 86
 4.2.1 Redis 事务的常用命令 86
 4.2.2 简单事务控制 87

4.2.3 取消一个事务 ………………… 87
4.2.4 乐观锁控制复杂事务 ………… 88
4.3 Redis 发布和订阅 …………………… 90
4.3.1 Redis 发布和订阅的
常用命令 ……………………… 90
4.3.2 Redis 发布和订阅实例 ……… 90
4.4 Redis 管道 …………………………… 91
4.5 数据备份与恢复 ……………………… 92
4.6 Redis 性能测试 ……………………… 93
4.7 Redis 客户端连接 …………………… 94
4.8 Redis 服务开机自启动 ……………… 94
4.8.1 Windows 下 Redis 服务
开机自启动 …………………… 95
4.8.2 Linux 下 Redis 服务
开机自启动 …………………… 96
4.9 Redis 内存分析工具 ………………… 97

第 5 章 Redis 缓存的持久化 ……………… 99
5.1 持久化机制 …………………………… 99
5.1.1 配置 RDB …………………… 100
5.1.2 配置 AOF …………………… 101
5.2 Redis 过期 key 清除策略 …………… 103

第 6 章 Redis 集群环境部署 ……………… 105
6.1 主从复制 ……………………………… 105
6.1.1 Redis 主从复制原理 ………… 106
6.1.2 Redis 主从复制安装过程 …… 106
6.1.3 Redis 测试主从复制关系 …… 108
6.2 哨兵模式 ……………………………… 109
6.2.1 灾备切换 Sentinel 的使用 …… 109
6.2.2 Redis Sentinel 的安装与
配置 …………………………… 111
6.2.3 测试主从切换 ………………… 116
6.3 Redis 集群 …………………………… 118
6.3.1 Redis 集群环境 ……………… 118
6.3.2 开始 Redis 集群搭建 ………… 119
6.3.3 Redis 集群代理 ……………… 125
6.3.4 Redis 集群特点 ……………… 128
6.3.5 新增 Redis 集群节点 ………… 130
6.3.6 删除 Redis 集群节点 ………… 139

第 7 章 Redis 开发实战 …………………… 142
7.1 搭建开发 Redis 的 Java
开发环境 ……………………………… 142
7.1.1 在 Windows 下安装 Java 8 …… 142
7.1.2 安装 Tomcat 9 ………………… 145
7.1.3 搭建 IntelliJ IDEA
开发环境 ……………………… 146
7.2 使用 Java 操作 Redis ………………… 147
7.2.1 连接 Redis 的两种方式 ……… 147
7.2.2 操作 String …………………… 150
7.2.3 操作 Map ……………………… 151
7.2.4 操作 List ……………………… 152
7.2.5 操作 Set ……………………… 153
7.2.6 排序 …………………………… 153
7.2.7 Redis 存储图片 ……………… 154
7.2.8 Redis 存储 Object …………… 158
7.2.9 Redis 存储和计算用户
访问量 ………………………… 161
7.3 Redis 调用方式 ……………………… 162
7.3.1 普通同步 ……………………… 162
7.3.2 事务 …………………………… 162
7.3.3 管道 …………………………… 163
7.3.4 管道中调用事务 ……………… 163
7.4 Redis 集群与 Java …………………… 164
7.5 实例 1：使用 Redis 获取用户的
共同好友 ……………………………… 164
7.5.1 初始化数据 …………………… 165
7.5.2 使用 Jedis 获取用户的
共同好友 ……………………… 165
7.6 实例 2：在 Tomcat 上使用 Redis
保存 Session ………………………… 166
7.6.1 分布式 Session ……………… 166
7.6.2 持久化 Tomcat Session 到
Redis ………………………… 167
7.6.3 安装服务器 Tomcat 和反向
代理服务器 Nginx …………… 168
7.6.4 配置 Tomcat 集群 …………… 173
7.6.5 配置 Tomcat 使用 Redis
管理 Session ………………… 177

第 8 章　Spring Boot 与 Redis 整合应用 179
8.1　Spring Boot 项目搭建与 Redis 整合应用 179
8.1.1　Spring Boot 简介 179
8.1.2　使用 Spring Initializr 新建项目 180
8.1.3　Spring Boot 结合 Redis 实战 184
8.2　RedisTemplate API 详解 186
8.2.1　写入和读取缓存 186
8.2.2　添加和获取散列数据 187
8.2.3　添加和获取列表数据 188
8.2.4　添加和获取集合数据 188
8.2.5　添加和获取有序集合数据 188
8.2.6　优化控制器 189
8.3　Spring Boot 集成 Spring Session 190
8.3.1　配置 Spring Boot 项目 190
8.3.2　创建配置类和控制器类 191
8.3.3　编译和部署项目 193

第 9 章　Redis 监控 196
9.1　Redis 监控指标 196
9.1.1　使用 INFO 命令 196
9.1.2　使用 redis-stat 206
9.2　自定义监控 208
9.2.1　前端页面 210
9.2.2　WebSocket 与消息推送 215
9.2.3　创建控制器类 220
9.2.4　业务逻辑 222
9.2.5　常用工具类 226

第 10 章　Redis 的缓存设计与优化 230
10.1　Redis 缓存的优点和缺点 230
10.2　缓存雪崩 231
10.3　缓存穿透 231
10.4　布隆过滤器 232
10.4.1　布隆过滤器简介 232
10.4.2　Redis 加载布隆过滤器模块 233
10.4.3　在项目中使用布隆过滤器 236

第 11 章　扩展知识 237
11.1　配置 CentOS 7 237
11.1.1　关闭防火墙 237
11.1.2　配置国内 yum 仓库 238
11.2　Maven 基础知识 239
11.2.1　Maven 的基本概念 239
11.2.2　Maven 下载 240
11.2.3　Maven 安装 241
11.2.4　修改从 Maven 中心仓库下载到本地的 JAR 包的默认存储位置 242
11.2.5　Maven 的简单使用 244
11.2.6　pom.xml 文件中的 groupId 和 artifactId 到底该怎么定义 244
11.2.7　常用 Maven 命令 244
11.3　配置 IntelliJ IDEA 245
11.3.1　配置 JDK 245
11.3.2　配置 Maven 247
11.3.3　配置 Tomcat 249
11.3.4　创建简单的 Maven 项目 250
11.3.5　导入 Maven 项目进行配置 253
11.4　使用 VMware 254
11.4.1　配置虚拟机的静态 IP 地址 254
11.4.2　恢复网络设置 255
11.4.3　重新生成虚拟机网卡的 MAC 地址 255
11.5　配置 SecureCRT 256
11.5.1　设置打开的连接显示在一个页面 257
11.5.2　传输文件和下载文件 257
11.5.3　显示中文 258
11.6　Chrome 的常用技巧 259
11.6.1　打开开发者工具控制台 259

11.6.2 基本输出·······················259
11.6.3 Chrome 禁用缓存············260
11.7 使用 Python 3 操作 Redis
　　 集群······························260

11.7.1 在 Windows 下安装
　　　 Python 3 ······················260
11.7.2 在 Linux 下安装 Python 3····266
11.7.3 使用 Redis 模块··············267

第 1 章 初识 Redis

本章我们将正式开始 Redis 学习之旅，主要学习以下内容。
- Redis 快速入门。
- Redis 环境搭建。
- Redis 可视化工具。
- 搭建 Redis 集群环境。

1.1 Redis 快速入门

1.1.1 Redis 简介

在当今的网络社会我们可以很方便地通过第三方平台（例如微博、百度等）访问和抓取数据。随着互联网 Web 2.0 网站的兴起，用户在社交网站上产生了海量的数据，使用传统的 SQL 数据库已经无法满足需求了。SQL 数据库面对大规模和高并发的 Web 2.0 网站会暴露很多问题，NoSQL 数据库却可以很好地处理用户的大量数据。我们先来了解什么是 NoSQL 数据库。

NoSQL（Not Only SQL），直译为"不仅仅是 SQL"，是非关系型数据库存储的广义定义，是一种全新的数据库。NoSQL 数据库以键值对（key-value）形式存储数据，和传统的关系型数据库不一样，不一定遵循传统的关系型数据库的一些基本要求，例如不遵循 SQL 标准、事务和表结构等。NoSQL 数据库主要有以下特点：非关系型的、分布式的、开源的和水平可扩展的。

到目前为止，已经涌现出了很多 NoSQL 数据库产品，例如 Redis、Memcached、MongoDB、Apache Cassandra、Apache CouchDB 等。国内的新浪微博也已经在其产品线上广泛地使用 Redis 作为其 NoSQL 数据库。

远程字典服务（Remote Dictionary Server，Redis）是一个开源的使用 ANSI C 编写、支持网络、可基于内存亦可持久化的日志型、key-value 数据库，它提供多种语言的 API，包括 Java、Python、PHP、C 和 C++等。Redis 的特点是高性能，可以适应高并发的应用场景和持久化存储。因此，可以说 Redis 纯粹为应用而生，是一个高性能的 key-value 数据库。

Redis 和另外一个非常流行的 NoSQL 数据库 Memcached 类似，但它支持存储的数据类型相对更多，包括字符串（String）、散列（Hash）、列表（List）、集合（Set）、有序集合（Sorted Set）。这些数据类型都支持 push/pop、add/remove 等命令，而且这些命令都是原子性（Atomic）的。此外，Redis 还支持各种不同方式的排序。

Redis 为了保证效率，把所有数据都保存在内存中，然后不定期地通过异步方式将数据保存到磁盘上（称为"半持久化模式"），也可以把每一次数据变化都写入 Appendonly 文件中（称为"全持久化模式"）。

Redis 较新版本的安装包是 Redis 6.0.6（截至 2020 年 8 月），可以从 Redis 官网下载，如图 1-1 所示。

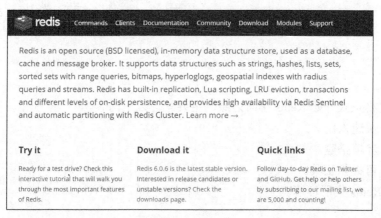

图 1-1 Redis 官网

我们也可以访问 Redis 的中文官网，如图 1-2 所示。这个页面有个方便的工具——互动教程，单击链接会弹出一个新的页面，如图 1-3 所示。这是一个网页版的、交互式的 Redis 环境，可以在这个页面中输入常用的 Redis 命令，适合新手练习使用。

图 1-2 Redis 的中文官网

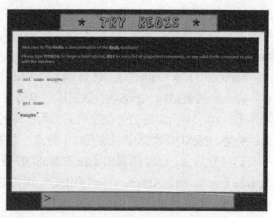

图 1-3 Redis 网页环境

1.1.2 Redis 特性

Redis 具有以下几个特性。

1. 丰富的数据类型

Redis 支持的 5 种基本数据类型如下。

- String。
- Hash。
- List。
- Set。
- Sorted Set。

这 5 种基本数据类型功能强大，在第 2 章会进行详细的介绍。

2．内存存储与持久化

Redis 把所有数据都存储在内存中。因为基于内存的读写速度快于硬盘，所以 Redis 在性能上比其他基于硬盘存储的数据库有更大的优势。

将数据存储在内存中也会出现问题，例如程序退出后内存中的数据会永久丢失。为此 Redis 提供对内存中数据的持久化支持，可以将内存中的数据异步写入硬盘中。

3．可用于缓存

Redis 中的消息可以通过 key 设置过期时间（Set Expire），过期后自动删除 key。Redis 使用 SETEX 命令来设置过期时间，按过期时间自动删除 key。

4．功能丰富

Redis 作为数据库，提供了丰富的功能，可以将它用作缓存、队列系统等。

Redis 可以为每个 key 设置过期时间，过期时间到期后 key 会自动被删除。这一功能配合 Redis 出色的性能让 Redis 可以作为缓存系统来使用，而且由于 Redis 支持持久化方式 RDB（它是 Redis Database 的缩写，原理是 Redis 在指定的时间间隔将内存中的数据库数据集快照写入磁盘）和 AOF（它是 Append Only File 的缩写，原理是将 Redis 的操作日志以追加的方式写入文件），因此 Redis 成了另一个非常流行的缓存系统 Memcached 的有力竞争者。

Redis 实现了 Master/Slave 这种主从同步的形式。Redis 还支持订阅模式，可以用来开发聊天室。

1.2 Redis 环境搭建

本节讲解如何在常见的操作系统上搭建和配置 Redis 的环境。读者可以从 Redis 官网下载最新版本的 Redis 安装包。

Redis 版本号采用标准惯例：主版本号.副版本号.补丁级别。偶数的副版本号用来表示稳定版本，如 1.2、2.0、2.2、2.4、2.6、2.8 等；奇数的副版本号用来表示非稳定版本，如 Redis 2.9.x 是非稳定版本，而 Redis 3.0 是稳定版本。

目前，Redis 在 Linux 下的稳定版本是 6.0.6，所以本书也选用这个稳定版本进行讲解。为了方便读者，本书搭建 Redis 环境所使用的 Redis 安装包和相关代码都托管到了 GitHub，安装详细情况和下载地址请参见笔者的 GitHub 个人主页。

1.2.1 在 Windows 下安装 Redis

本小节讲解在 Windows 下安装并配置 Redis 环境。Redis 的 Windows 安装环境信息如表 1-1 所示。

表 1-1　　　　　　　　　　Redis 的 Windows 安装环境信息

操作系统	Windows 10 64 位
Redis-x64	3.2.100

1．基于 Windows 64 位的 Redis 安装

Redis 没有官方的 Windows 版本，但是微软开源技术团队（Microsoft Open Tech Group）长期开发和维护着这个基于 Windows 64 位的 Redis 版本，更多详细信息参考 GitHub 上的项目主页。进入主页，会发现有以下信息需要注意，如图 1-4 所示。

Redis on Windows

- This is a port for Windows based on Redis.
- We officially support the 64-bit version only. Although you can build the 32-bit version from source if desired.
- You can download the latest unsigned binaries and the unsigned MSI installer from the release page.
- For releases prior to 2.8.17.1, the binaries can found in a zip file inside the source archive, under the bin/release folder.
- Signed binaries are available through NuGet and Chocolatey.
- Redis can be installed as a Windows Service.

图 1-4　需要注意的信息

Windows 下的 Redis 具有以下特点。
- Redis 服务器使用一个默认端口（6379）。
- 微软开源技术团队只正式支持 64 位的 Redis 版本。如果需要的话，可以从源代码构建 32 位的 Redis 版本。
- 可以从发布页面下载最新的未签名的二进制文件和未签名的 MSI 安装程序。
- 对于 Redis 2.8.17.1 之前的版本，其二进制文件可以在源文件中的 bin/release 目录下的扩展名为.zip 的文件中找到。
- 签名的二进制文件可通过 NuGet 和 Chocolatey 获得。
- Redis 可以作为 Windows 服务进行安装。

本小节中使用的是基于 Windows 64 位的 Redis 安装包。

访问 Redis 的官方下载地址，找到 Windows 64 位的 Redis 安装包 Redis-x64-3.2.100.zip，如图 1-5 所示。

图 1-5　下载 Windows 64 位的 Redis 安装包

Redis 的 Windows 开源版本只支持 Windows 64 位。下载 Redis-x64-3.2.100.zip 安装包到 D 盘，解压缩后将文件夹重新命名为 Redis，可以看到 Redis 文件夹中有如下文件，如图 1-6 所示。

Redis 的可执行命令和配置文件说明如下。

- redis-benchmark.exe：测试工具，测试 Redis 的读写性能情况。
- redis-check-aof.exe：用于数据导入，AOF 文件修复工具。
- redis-cli.exe：Redis 客户端（Client）程序。
- redis-server.exe：Redis 服务器（Server）程序。
- redis.windows.conf：Redis 在 Windows 下的配置文件，主要是一些 Redis 的默认服务配置，包括默认端口号（6379）等。

图 1-6　Redis 文件夹中的文件

2．启动 Redis 服务器

在 Windows 下按 "Win + R" 组合键执行 cmd 命令，进入命令提示符窗口，cmd 是 command 的缩写，是 Windows 下的一个命令。命令提示符窗口，如图 1-7 所示。

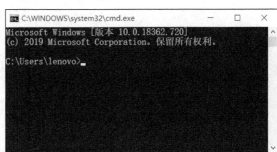

图 1-7　命令提示符窗口

使用 cd 命令切换到 D:\Redis 目录下，再执行以下命令。

```
redis-server.exe Redis.windows.conf
```

执行命令后，如果在窗口中显示图 1-8 所示的信息，那么 Redis 服务器就启动成功了，在窗口提示中会带有 Redis 服务器的版本号、运行进程号（PID）、运行端口信息（Port），默认的监听端口是 6379。

执行 redis-server 命令会启动 Redis 服务，是在前台控制台中直接运行的。也就是说，执行完该命令后，如果关闭当前命令提示符窗口，Redis 服务也会关闭。因此一般会将其改为从后台的 Windows 服务启动，并且设置为开机自动启动，就像数据库服务器中的 SQL Server 服务和 Web 服务器中的 IIS 服务一样。这部分内容详见 4.8.1 小节 Windows 下 Redis 服务开机自启动。

图1-8　Redis服务器启动成功

还需要把 Redis 的安装目录添加到系统变量 Path 中。在桌面上右键单击"此电脑",弹出快捷菜单,选择"属性"→"高级系统设置"→"高级",单击"环境变量"按钮,弹出"环境变量"对话框,如图1-9所示。

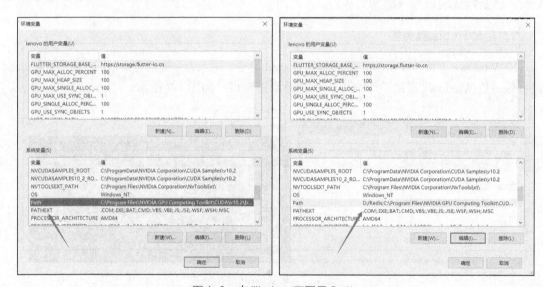

图1-9　在 Windows 下配置 Redis

在系统变量 Path 中添加以下变量值,也就是 Redis 的安装目录。
D:/Redis;

（1）D:/Redis 是笔者在本机上安装 Redis 的位置,读者需要根据自己计算机上的实际情况进行修改。
（2）D:/Redis 是安装目录的路径,为了避免字符串转义,路径分隔符使用"/"。
（3）Windows 系统变量 Path 中的变量之间使用";"进行分隔。

3．启动 Redis 客户端

重新启动一个新的命令提示符窗口,原来的命令提示符窗口不要关闭,不然就无法访问

Redis 服务器了。在新的命令提示符窗口中执行以下命令启动 Redis 客户端。

redis-cli.exe -h 127.0.0.1 -p 6379

客户端的基本参数如下。

-h：设置检测主机 IP 地址，默认为 127.0.0.1

-p：设置检测主机的端口号，默认为 6379

每次启动 Redis 客户端都需要输入 redis-cli 命令，这样操作不是很方便。在 Windows 下，我们可以新建一个批处理文件，命名为 runRedisClient.bat，把启动 Redis 客户端的命令写在这个批处理文件里，以后只要单击这个批处理文件就可以启动 Redis 客户端了，如图 1-10 所示。该批处理文件存放在 Redis/Chapter01 目录下。

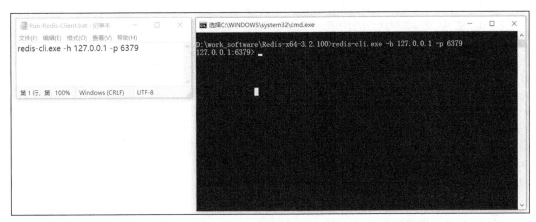

图 1-10　启动 Redis 客户端

4．关闭 Redis 服务器

在 Redis 客户端执行 shutdown 命令，会关闭 Redis 服务器（如图 1-11 所示），然后持久化 Redis 内存中的数据到文件中。

图 1-11　关闭 Redis 服务器

1.2.2　在 Linux 下安装 Redis

本小节我们将在 CentOS 7 上通过 root 账号安装 Redis，编译源代码，安装二进制文件，创建和安装文件。CentOS（Community Enterprise Operating System）是 Linux 发行版之一，它由 Red Hat Enterprise Linux（RHEL）发布的源代码编译而成。由于出自同样的源代码，因此有些要求高度稳定性的服务器以 CentOS 代替商业版的 Red Hat Enterprise Linux。

首先，从 Redis 官网下载最新的稳定版本的 Redis 源码包，本书使用的最新的稳定版本是 Redis 6.0.6，如图 1-12 所示。

Redis 6 开发与实战

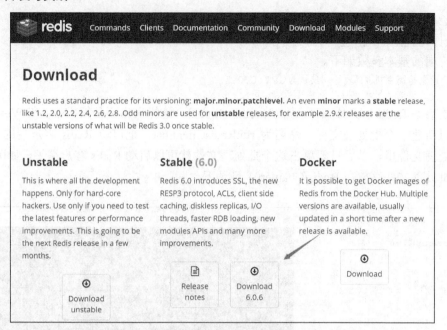

图1-12　下载最新的稳定版本的 Redis 源码包

Redis 的 Linux 安装环境信息如表1-2所示。

表1-2　Redis 的 Linux 安装环境信息

操作系统	CentOS 7 64 位
Redis	6.0.6
gcc	9.3.1

其次，安装 gcc 基础依赖包，使用如下命令。

```
$ yum -y install gcc-c++
```

再次，为了编译最新版本的 Redis 源码还需要使用 devtoolset 升级 gcc 版本。

```
$ yum -y install centos-release-scl
$ yum -y install devtoolset-9-gcc devtoolset-9-gcc-c++ devtoolset-9-binutils
$ scl enable devtoolset-9 bash
$ echo "source /opt/rh/devtoolset-9/enable" >>/etc/profile
```

安装 devtoolset 后，需要输入 scl enable devtoolset-9 bash 来启动 devtoolset。启动 devtoolset 后仅针对本次会话有效，若重新登录 Linux，需要再次使用 scl 命令激活 devtoolset。

若要使 devtoolset 长期有效，需要输入 echo "source /opt/rh/devtoolset-9/enable">>/etc/profile。

然后，输入 gcc -v 命令查看升级后的 gcc 版本，如图1-13所示，可以看出当前的 gcc 版本为 9.3.1，使用这个版本的 gcc 来编译 Redis 源码。

```
[root@localhost ~]# gcc -v
Using built-in specs.
COLLECT_GCC=gcc
COLLECT_LTO_WRAPPER=/opt/rh/devtoolset-9/root/usr/libexec/gcc/x86_64-redhat-linux/9/lto-wrapper
Target: x86_64-redhat-linux
Configured with: ../configure --enable-bootstrap --enable-languages=c,c++,fortran,lto --prefix=/opt/rh/devtoolset-9/root/u
sr --mandir=/opt/rh/devtoolset-9/root/usr/share/man --infodir=/opt/rh/devtoolset-9/root/usr/share/info --with-bugurl=http:
//bugzilla.redhat.com/bugzilla --enable-shared --enable-threads=posix --enable-checking=release --enable-multilib --with-s
ystem-zlib --enable-__cxa_atexit --disable-libunwind-exceptions --enable-gnu-unique-object --enable-linker-build-id --with
-gcc-major-version-only --with-linker-hash-style=gnu --with-default-libstdcxx-abi=gcc4-compatible --enable-plugin --enable
-initfini-array --with-isl=/builddir/build/BUILD/gcc-9.3.1-20200408/obj-x86_64-redhat-linux/isl-install --disable-libmpx -
-enable-gnu-indirect-function --with-tune=generic --with-arch_32=x86-64 --build=x86_64-redhat-linux
Thread model: posix
gcc version 9.3.1 20200408 (Red Hat 9.3.1-2) (GCC)
```

图1-13　查看 gcc 版本

接着，建立 Redis 下载目录/upload，在这个目录里下载最新的稳定版本的 Redis 源码包。

```
$ mkdir /upload
$ cd /upload
$ wget http://download.redis.io/releases/redis-6.0.6.tar.gz
```

解压缩 Redis 源码包。

```
$ tar -xzvf redis-5.0.5.tar.gz
```

进入解压缩后的目录。

```
$ cd redis-6.0.6/
```

使用 make 命令编译 Redis 源文件。

```
$ make
```

编译成功后，安装 Redis。

```
$ make install PREFIX=/usr/local/redis
```

注意　　make 命令使用的参数 PREFIX 代表安装路径，参数要大写。

执行 make 命令后会自动把 Redis 的可执行命令复制到/usr/local/redis/bin 目录下，这样执行 Redis 命令时，就不用输入完整路径了。进入安装目录 bin 下，此时 bin 目录下的文件如图 1-14 所示。

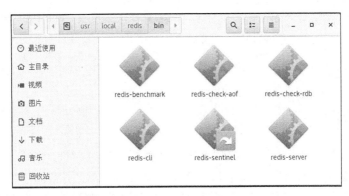

图 1-14　Redis 的 bin 目录下的文件

Redis 的 bin 目录下的文件说明如下。

- redis-benchmark：Redis 性能测试工具。
- redis-check-aof：文件修复工具。
- redis-check-rdb：文件修复工具。
- redis-cli：Redis 命令行客户端。
- redis-sentinel：Redis 集群管理工具。
- redis-server：Redis 服务进程命令。

安装成功后，需要对 Redis 进行部署，把 Redis 的配置文件 redis.conf 复制到 /usr/local/redis/conf 目录下。

```
$ mkdir -p /usr/local/redis/conf
$ cp redis.conf /usr/local/redis/conf
```

这时在/usr/local/redis/bin 目录下就包含 Redis 所有可执行命令，在/usr/local/redis/conf 目录下放置 redis 的配置文件 redis.conf。部署后的 Redis 目录结构如图 1-15 所示。

最后，将 Redis 的可执行命令所在目录添加到系统变量 Path 中，修改/etc/profile 文件。

$ vi /etc/profile

在/etc/profile 文件的最后一行添加以下内容。

export PATH=$PATH:/usr/local/redis/bin

然后输入 source 命令使这个文件立即生效。

$ source /etc/profile

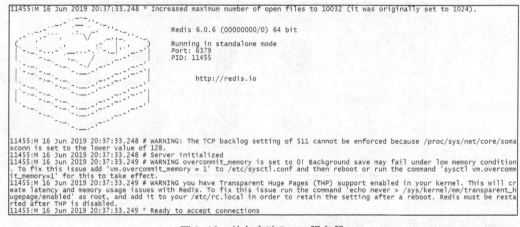

图 1-15 部署后的 Redis 目录结构

至此，Redis 在 Linux 上的安装和配置就结束了。

1. 启动 Redis 服务器并加载指定配置文件

启动 Redis 服务器分为前台启动 Redis 服务器和后台启动 Redis 服务器。前台启动 Redis 服务器将占用当前命令行窗口。若从前台启动 Redis 服务器，执行 redis-server 命令时需要指定配置文件 redis.conf。

$ redis-server /usr/local/redis/conf/redis.conf

前台启动 Redis 服务器成功后如图 1-16 所示。这个页面将占用命令行窗口，退出请按 "Ctrl + C" 组合键。

图 1-16 前台启动 Redis 服务器

这里直接执行 redis-server 命令来启动 Redis 服务器，是在前台直接运行的（效果如图 1-15 所示）。也就是说，执行完该命令后，如果 Linux 关闭当前会话，则 Redis 服务也随即关闭。

从后台启动 Redis 服务器，需要指定启动 Redis 的配置文件 redis.conf。这部分内容详见 4.8.2 小节 Linux 下 Redis 服务开机自启动。

如果要从后台启动 Redis 服务器，还需要修改配置文件 redis.conf 里的 daemonize 对应的参数为 yes。

2. 查看 Redis 状态

启动 Redis 服务器后可以使用图 1-17 所示命令查看 Redis 的进程。

第1章 初识Redis

图1-17 使用ps命令查看Redis的进程

如图1-17所示，Redis的进程号是73459，可以执行 kill -9 <PID> 命令关闭Redis进程。

```
$ kill -9 73459
```

也可以在控制台执行 redis-cli shutdown 命令关闭Redis。

```
$ redis-cli shutdown
```

3. 启动Redis客户端

Redis服务器启动以后，就打开另一个客户端控制台（Terminal），输入redis-cli命令进行测试，检验是否成功连接本机的Redis服务器，如图1-18所示。

图1-18 客户端连接Redis服务器

如图1-18所示，说明Redis服务器已经正常工作，如果Redis服务器未启动，则执行redis-cli命令时会报 Could not connect to Redis at 127.0.0.1:6379: Connection refused 错误。

4. 设置RDB快照地址

Redis提供了将内存中的数据持久化到硬盘，以及用持久化文件来恢复数据库数据的功能，Redis支持两种形式的持久化，一种是RDB快照（snapshotting），另一种是AOF（append-only-file），这部分内容在第5章 Redis 缓存持久化会详细解释。

其中RDB快照是把当前内存中的数据集快照写入磁盘，也就是快照（数据库中所有键值对数据）。快照的文件名默认是dump.rdb。这个dump.rdb文件存放位置是不固定的，默认存放在启动Redis服务器时的当前目录，有两种方法修改dump.rdb的存放位置。

方法一：使用CONFIG SET dir命令来手工指定dump.rdb的存放目录。

```
# 修改dump.rdb文件的存放路径为/usr/local/redis/conf
127.0.0.1:6379> CONFIG SET dir /usr/local/redis/conf
OK
127.0.0.1:6379> CONFIG GET dir
1) "dir"
2) "/usr/local/redis/conf"
```

方法二：修改redis.conf配置文件里的dir配置项。

```
dir /usr/local/redis/conf
```

然后输入redis-server /usr/local/redis/conf/redis.conf，使用redis-server命令重新加载redis.conf配置文件，使Redis的配置文件生效。

5. 开启 Redis 多线程

Redis 6 加入了多线程功能。使用 Redis 6 来处理网络数据的读写和协议解析，执行命令依然是以单线程的形式，要开启 Redis 的多线程功能，可以在 redis.conf 配置文件中加入以下配置项。

```
io-threads-do-reads yes      # 开启 IO 线程
io-threads 3                 # 设置 IO 线程数
```

io-threads 线程数最多可设置为 Linux 服务器中 CPU 核心数的 3/4，也就是说如果 CPU 有 4 个核心，尝试把这个值设置为 3，如果 CPU 有 8 个核心，尝试把这个值设置为 6。但不建议这个值超过 8，当设置多于 8 个线程时，不会有明显的性能提升。查看 Linux 服务器 CPU 核心数的命令是 lscpu，如图 1-19 所示。

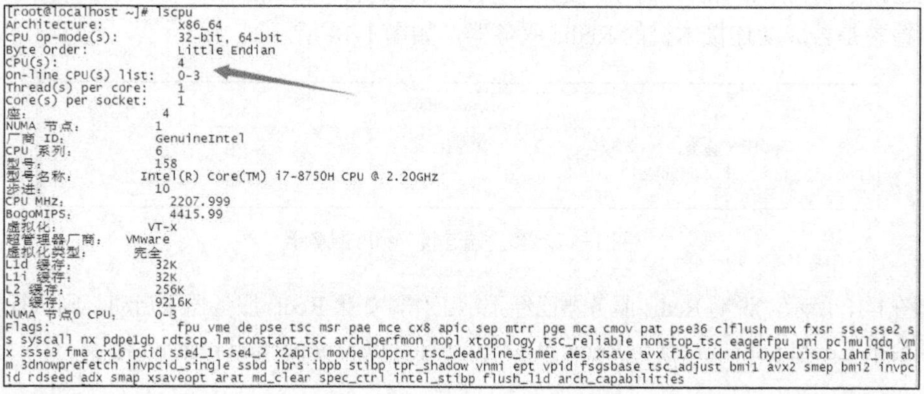

图 1-19 查看 CPU 核心数

可以看出 CPU 核心数为 4，所以 io-threads 配置项设置为 3，读者需要根据自己机器的实际情况进行修改。

6. 关闭 Redis 服务器

在新的客户端控制台执行 shutdown 命令，会关闭 Redis 服务器，如图 1-20 所示。

图 1-20 关闭 Redis 服务器

最后会显示图 1-21 所示的提示。

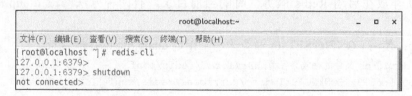

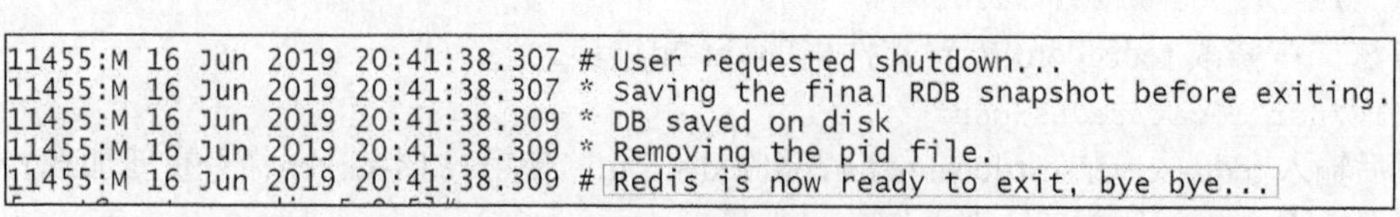

图 1-21 Redis 服务器关闭时的提示

1.3 Redis 可视化工具

Redis 与传统的关系型数据库一样，也有图形化的管理工具：Redis Desktop Manager。Redis Desktop Manager 是一款好用的 Redis 桌面管理工具，支持命令控制台操作，以及常用查询 key、rename、delete 等操作。读者可在 Redis Desktop Manager 官网下载 Redis Desktop Manager。

Redis Desktop Manager 启动界面如图 1-22 所示。

图 1-22 Redis Desktop Manager 启动界面

单击"Connect to Redis Server"按钮来连接 Redis 服务器。在弹出的"Connection"对话框中输入连接名称、Redis 服务器的 IP 地址、端口和连接密码，如图 1-23 所示。

图 1-23 输入连接 Redis 的信息

按照图 1-23 所示的样式创建一个连接，打开 Redis 非关系型数据库后的界面如图 1-24 所示。

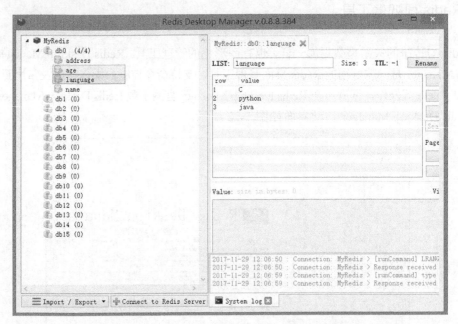

图 1-24　打开 Redis 非关系型数据库后的界面

1.4　搭建 Redis 集群环境

计算机集群由一组连接在一起的计算机组成，计算机集群将每个节点（用作服务器的计算机）设置为执行相同的任务，由软件控制和调度。部署集群是为了提高计算机系统的性能和可用性。

搭建 Redis 集群环境至少需要 6 个节点，包括 3 个 Master（主节点）和 3 个 Slave（从节点），每个节点之间都可以相互通信。Redis 集群环境是在 Linux 下进行搭建的。为了完成第 6 章的 Redis 集群环境部署，本书使用虚拟机软件搭建了 6 个独立的 Linux 虚拟机，组成一个完整的网络环境。

本书的实验环境是在 Windows 10 下使用 VMware 虚拟机软件来安装 Linux，使用的相关软件如下。

- 虚拟机软件：VMware Workstation 15 Pro。
- Linux：CentOS Linux 7.4(1708) released。
- 终端仿真软件：SecureCRT 7.0.0。
- 文件上传工具：SecureFX 7.0.0。

虚拟机软件是一种特殊的软件，可以在计算机和终端用户之间创建一种环境，而终端用户基于虚拟机软件所创建的环境来操作其他软件。例如，可以实现一台计算机同时运行多个操作系统，这些操作系统可以是 Windows、Linux 和 macOS 等虚拟机软件支持的操作系统。本书使用的虚拟机软件是 VMware。

一般把安装虚拟机软件的操作系统称为宿主机，虚拟机软件上安装的操作系统称为虚拟机。当设置虚拟机的网络选项后，就可以将宿主机与虚拟机联通，使宿主机上的几个虚拟机组成一个互联互通的局域网。

1.4.1 配置 VMware 准备安装 CentOS

启动 VMware 后，单击"创建新的虚拟机"，如图 1-25 所示。

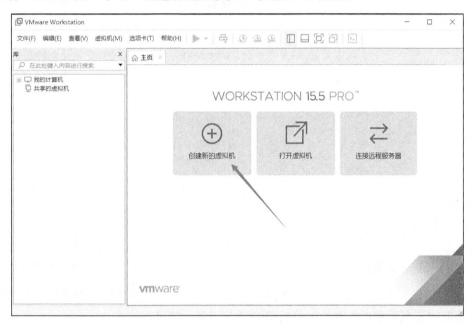

图 1-25　创建新的虚拟机

在弹出的"新建虚拟机向导"对话框中选择"自定义（高级）"，然后单击"下一步"按钮，如图 1-26 所示。

在"新建虚拟机向导"的硬件兼容性对应的下拉列表框中选择"Workstation 12.0"，然后单击"下一步"按钮，如图 1-27 所示。

图 1-26　虚拟机自定义安装

图 1-27　选择虚拟机硬件兼容性

在"新建虚拟机向导"对话框中,选择"稍后安装操作系统",然后单击"下一步"按钮,如图1-28所示。

在"新建虚拟机向导"对话框中,选择客户机操作系统为"Linux",选择版本为"CentOS 64位",然后单击"下一步"按钮,如图1-29所示。

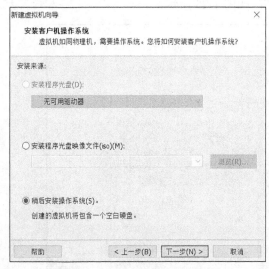

图1-28 安装客户机操作系统

图1-29 选择客户机操作系统

在图1-30所示对话框中填入虚拟机名称,在"位置"文本框中输入合适的路径来保存在虚拟机上创建磁盘后形成的文件,然后单击"下一步"按钮。

在图1-31所示对话框中根据自己计算机上的CPU个数选择处理器数量和每个处理器的核心数量,然后单击"下一步"按钮。

图1-30 命名虚拟机

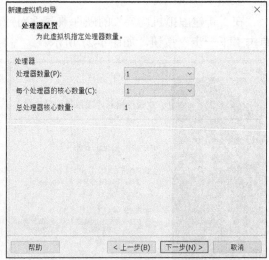

图1-31 处理器配置

在"新建虚拟机向导"对话框中为虚拟机设置合适的内存,根据计算机的配置选择内存大小。为了顺利进行Redis集群实验,内存建议不低于512MB,这要求宿主机的物理内存不低于 8GB，这里我们设置为 1024MB，然后单击"下一步"按钮，如图 1-32 所示。

在图 1-33 所示对话框中设置虚拟机的网络类型，选择网络连接为"使用网络地址转换（NAT）"，然后单击"下一步"按钮。

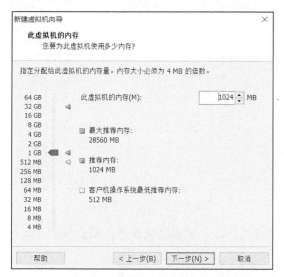

图 1-32　设置虚拟机的内存

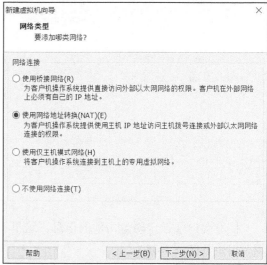

图 1-33　设置虚拟机的网络类型

在图 1-34 所示对话框中选择 SCSI 控制器为"LSI Logic"，然后单击"下一步"按钮。
在图 1-35 所示对话框中选择虚拟磁盘类型为"SCSI"，然后单击"下一步"按钮。

图 1-34　选择 I/O 控制器类型

图 1-35　选择磁盘类型

在图 1-36 所示对话框中选择磁盘为"创建新虚拟磁盘"，然后单击"下一步"按钮。

VMware 针对安装 CentOS 64 位虚拟机的磁盘容量建议大小是 20GB，设置磁盘容量时大小要合适，为了以后的 Redis 集群实验，这里设置为 50GB。选择"将虚拟磁盘存储为单个文件"，然后单击"下一步"按钮，如图 1-37 所示。

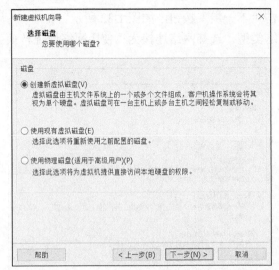

图 1-36 选择磁盘

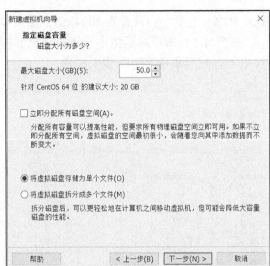

图 1-37 指定磁盘容量

接着会出现指定磁盘文件的信息，这里不做任何修改，采用默认设置，然后单击"下一步"按钮，如图 1-38 所示。

之后会出现已准备好创建虚拟机的信息，表示虚拟机安装完成，如图 1-39 所示。

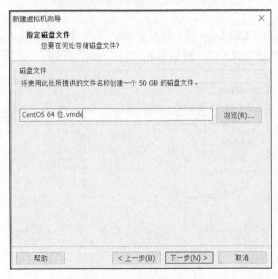

图 1-38 指定磁盘文件

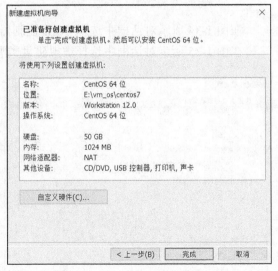

图 1-39 已准备好创建虚拟机

1.4.2 安装 Linux

配置完虚拟机软件后，接下来就可以基于 VMware 虚拟机软件安装 CentOS 了。在 CentOS 官网单击 "Get CentOS Now" 按钮下载 CentOS 7，如图 1-40 所示。下载到本地硬盘的 CentOS 安装文件是 CentOS-7-x86_64-Everything-1810.iso 镜像文件。

安装 CentOS，首先单击虚拟机的 "CD/DVD(IDE)"，如图 1-41 所示。在弹出的 "虚拟机设置"对话框中选择"使用 ISO 映像文件"，然后单击"浏览"按钮，选择之前下载好的 CentOS 安装文件（ISO 格式），如图 1-42 所示。

图 1-40　CentOS 官网

图 1-41　安装 CentOS

图 1-42　虚拟机设置

配置虚拟机完成后，单击 VMware 工具栏上的"Power On"按钮（绿色的三角形），启动虚拟机，开始安装 CentOS，如图 1-43 所示。之后的过程与在硬盘上安装 CentOS 相同，由于本书篇幅所限，不再详述。

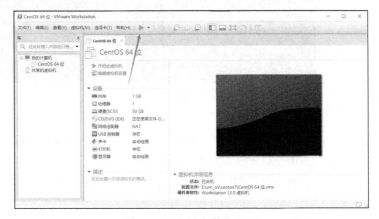

图 1-43　开始安装 CentOS

1.4.3 安装 VMware Tools

在 VMware 下安装 CentOS 完毕后，可以安装 VMware 的 VMware Tools 来提高鼠标操作和屏幕显示性能。安装 VMware Tools 可以使虚拟机和宿主机互相传输文件，可以在虚拟机全屏后让 CentOS 扩展到全屏显示，如图 1-44 所示。

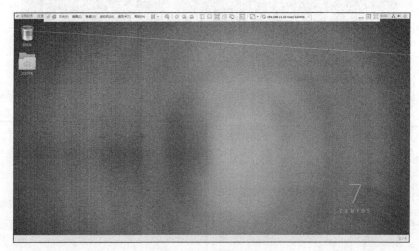

图 1-44　全屏显示 CentOS

启动虚拟机中的 CentOS。CentOS 启动完毕后，在 VMware 的"虚拟机"菜单中选择"安装 VMware Tools"，如图 1-45 所示。之后会自动在虚拟光驱装载 VMware Tools 软件包，如图 1-46 所示。

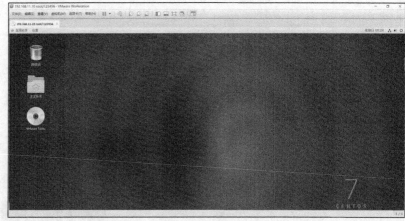

图 1-45　安装 VMware Tools　　　　　图 1-46　已安装 VMware Tools

双击虚拟机桌面上的光驱图标，这时显示光驱中有 5 个文件，如图 1-47 所示。复制其中的 VMwareTools-10.0.10-4301679.tar.gz 文件到桌面。

然后启动终端，使用 tar 命令解压缩安装文件，并执行 vmware-install.pl 安装 VMware Tools。

```
[root@localhost 桌面]# tar -zxvf VMwareTools-10.0.10-4301679.tar.gz
[root@localhost 桌面]# cd vmware-tools-distrib/
```

```
[root@localhost vmware-tools-distrib]# ll
总用量 504
drwxr-xr-x.  2 root root     87 8月  26 2016 bin
drwxr-xr-x.  5 root root     39 8月  26 2016 caf
drwxr-xr-x.  2 root root     67 8月  26 2016 doc
drwxr-xr-x.  5 root root   4096 8月  26 2016 etc
-rw-r--r--.  1 root root 284342 8月  26 2016 FILES
-rw-r--r--.  1 root root   2538 8月  26 2016 INSTALL
drwxr-xr-x.  2 root root    137 8月  26 2016 installer
drwxr-xr-x. 15 root root    202 8月  26 2016 lib
drwxr-xr-x.  3 root root     21 8月  26 2016 vgauth
-rwxr-xr-x.  1 root root    243 8月  26 2016 vmware-install.pl
-rwxr-xr-x.  1 root root 214600 8月  26 2016 vmware-install.real.pl
```

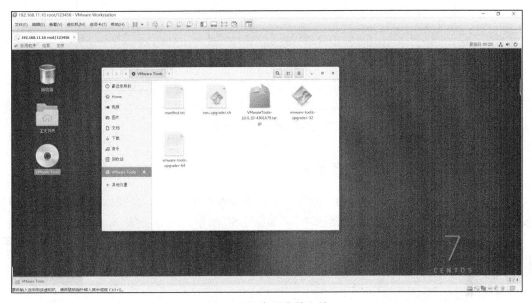

图 1-47 光驱中的文件

接下来执行 vmware-install.pl 这个 Perl 文件，执行过程的最后会根据显示器选择一个合适的屏幕分辨率。

```
[root@localhost vmware-tools-distrib]# sudo ./vmware-install.pl
```

安装 pl 脚本后，为了使配置生效需要重新启动 CentOS。

1.4.4 虚拟机与宿主机的网络设置

为了完成 Redis 集群实验，需要设置虚拟机的网络连接方式，使得虚拟机之间可以互联互通，组成一个小型的局域网。本小节中 VMware 使用 NAT 模式搭建内网和外网都可以访问的虚拟机。

1. 查看虚拟机服务是否开启

按"Win+R"组合键，在文本框中输入 services.msc，如图 1-48 所示。单击"确定"按钮，弹出"服务"窗口。

图 1-48 输入 services.msc

在"服务"窗口中查看以 VMware 开头的服务是否开启,没有开启的话就选中服务并右击,选择"启动"来开启服务,如图 1-49 所示。

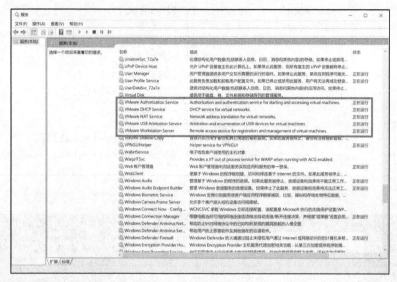

图 1-49　开启 VMware 服务

2．查看虚拟机的虚拟网卡是否启用

在桌面右下角网络标志处右击,选择"网络和 Internet 设置",在弹出的"设置"窗口中选择"以太网",单击"更改适配器选项",如图 1-50 所示。

查看虚拟机的虚拟网卡是否启用,没有启用的话就选中虚拟机网卡并右击,选择"启用"来启用虚拟机网卡,如图 1-51 所示。

图 1-50　网络和 Internet 设置

图 1-51　查看虚拟网卡是否启用

Vmware Network Adapter VMnet8 虚拟网卡启用后需要设置 IP 地址,本书 VMnet8 设置的 IP 地址为 192.168.11.1,如图 1-52 所示。

图 1-52　VMnet8 设置的 IP 地址

设置完成后，进入命令提示符窗口，执行 ipconfig 命令查看 VMnet8 的 IP 地址为 192.168.11.1，如图 1-53 所示。从图中可以看出宿主机 VMnet8 的 IP 地址 192.168.11.1 属于 11 网段。

3．设置虚拟机

打开虚拟机，选中需要操作的虚拟机并右击，选择"设置"选项，弹出"虚拟机设置"对话框。在"虚拟机设置"对话框中选择"网络适配器"，选择网络连接为"NAT 模式"，然后单击"确定"按钮完成设置，如图 1-54 所示。

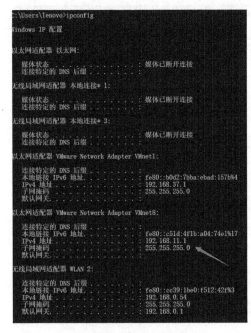

图 1-53　查看 IP 地址

图 1-54　设置虚拟机的网络连接

4. 设置虚拟网络

单击 VMware 虚拟机菜单栏的"编辑"菜单,选择"虚拟网络编辑器"选项,如图 1-55 所示。

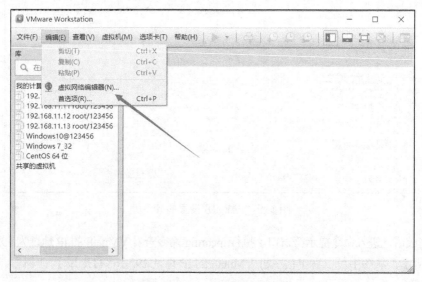

图 1-55 打开虚拟网络编辑器

打开"虚拟网络编辑器"对话框后,选择"VMnet8",单击"更改设置"按钮,如图 1-56 所示。

修改 VMnet8 的子网 IP 为 192.168.11.0,子网掩码为 255.255.255.0,如图 1-57 所示。

图 1-56 在"虚拟网络编辑器"对话框中更改设置

图 1-57 修改子网 IP 和子网掩码

然后单击"NAT 设置"按钮,弹出"NAT 设置"对话框,修改网关 IP 为 192.168.11.2,如图 1-58 所示。

第 1 章 初识 Redis

图 1-58　修改网关 IP

5．配置 CentOS 的静态 IP 地址

在 CentOS 的控制台中，修改/etc/sysconfig/network-scripts/ifcfg-ens33 文件的配置。

```
$ vi /etc/sysconfig/network-scripts/ifcfg-ens33
```

修改后的配置内容如下。

```
TYPE=Ethernet
BOOTPROTO=static
NAME=ens33
DEVICE=ens33
ONBOOT=yes
IPADDR=192.168.11.10        # CentOS 的静态 IP 地址
GATEWAY=192.168.11.2        # CentOS 的网关地址
NETMASK=255.255.255.0
DNS1=192.168.11.2
```

修改/etc/resolv.conf 文件。

```
$ vi /etc/resolv.conf
```

添加以下内容。

```
nameserver 192.168.11.2
```

修改完配置文件后，执行 service network restart 命令，重启网络，如图 1-59 所示。

1.4.5　复制虚拟机

为了方便使用虚拟机，减少重复配置，可以直接将配置好的虚拟机进行复制，复制虚拟机时被复制的虚拟机必须处于停机状态。在 1.4.2 小节创建的虚拟机 CentOS 7 位于 E:\vm_os\centos7 目录下，复制 centos7 文件夹并将其重新命名为 centos64-2，如图 1-60 所示。

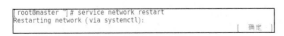

图 1-59　重启网络　　　　　　　　图 1-60　复制虚拟机

25

打开 E:\vm_os\centos64-2 目录下的 CentOS 64 位.vmx 文件，如图 1-61 所示。在 VMware 中加载复制的虚拟机，如图 1-62 所示。

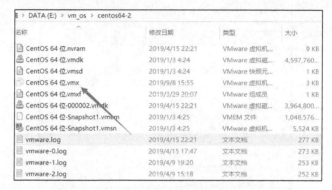

图 1-61　打开 CentOS 64 位.vmx 文件

图 1-62　加载复制的虚拟机

为了区分虚拟机，以虚拟机的 IP 地址重新命名虚拟机，如图 1-63 所示。

启动名为 192.168.11.11 的虚拟机时，会弹出提示对话框，如图 1-64 所示。选择第二个选项"我已复制该虚拟机"。

图 1-63　以 IP 地址命名的虚拟机

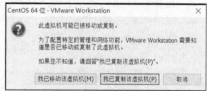

图 1-64　复制该虚拟机

复制虚拟机后，由于复制的虚拟机与被复制的虚拟机信息完全一致，因此需要将复制的虚拟机重新进行网络信息的设置，设置的步骤如下。

1. 设置 IP 地址为 192.168.11.11 的虚拟机的静态 IP 地址

修改 vi /etc/sysconfig/network-scripts/ifcfg-ens33 文件的配置如下。

```
TYPE=Ethernet
BOOTPROTO=static
NAME=ens33
DEVICE=ens33
```

```
ONBOOT=yes
IPADDR=192.168.11.11       # 虚拟机 CentOS 的 IP 地址
GATEWAY=192.168.11.2       # 虚拟机 CentOS 的网关地址
NETMASK=255.255.255.0
DNS1=192.168.11.2
```

2．设置虚拟机 192.168.11.11 的 DNS 客户端配置文件

修改/etc/resolv.conf 文件。

`vi /etc/resolv.conf`

添加以下内容。

`nameserver 192.168.11.2`

修改完配置文件后，执行 service network restart 命令，重启网络使配置文件生效，如图 1-65 所示。

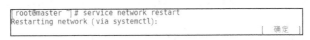

图 1-65　重启网络

到这里，虚拟机 192.168.11.11 的配置就修改完成了，按照同样的配置添加虚拟机 192.168.11.12、192.168.11.13、192.168.11.14 和 192.168.11.15，配置完成后如图 1-66 所示。

图 1-66　配置 Redis 集群环境

Redis 集群环境的 IP 地址和操作系统如表 1-3 所示。

表 1-3　　　　　　　　　Redis 集群环境的 IP 地址和操作系统

IP 地址	操作系统
192.168.11.10	CentOS 7 x86_64
192.168.11.11	
192.168.11.12	
192.168.11.13	
192.168.11.14	
192.168.11.15	

至此，Redis 的虚拟机集群环境搭建完毕。

第 2 章 Redis 常用数据类型及命令

Redis 的数据存储结构是 key-value 对，定义 key 时要注意以下几点。

（1）key 不要太长，尽量不要超过 1024Byte，太长的话不仅消耗内存，而且会降低查找的效率。

（2）key 也不要太短，太短的话，key 的可读性会降低。

（3）一个项目中，key 单词与单词之间以:分开，例如 SET user-name: loginnamewangwu。

本章主要介绍 Redis 的 5 种数据类型，即 String、Hash、List、Set 及 Sorted Set。

2.1 String 类型

String 类型是 Redis 最基本的数据类型，一个 key 对应一个 value。String 类型是二进制安全的，即 Redis 的 String 类型可以包含任何数据，例如扩展名为.jpg 的图片或者序列化的对象。这部分内容可以参考 7.2 节使用 Java 操作 Redis。

2.1.1 SET

SET 命令的基本语法如下。

```
SET key value
```

SET 用于将字符串值 value 关联到 key。如果 key 已经持有其他值，SET 就覆盖其他值。

返回值：总是返回 OK，因为 SET 不可能失败。

实例 1：对字符串类型的 key 执行 SET 命令。

```
127.0.0.1:6379> SET name xinping
OK
127.0.0.1:6379> GET name
"xinping"
```

实例 2：对非字符串类型的 key 执行 SET 命令。

```
127.0.0.1:6379> LPUSH greet_list "hello"    # 建立一个列表
(integer) 1

127.0.0.1:6379> TYPE greet_list
list

127.0.0.1:6379> SET greet_list "world"      # 覆盖列表类型
OK
```

```
127.0.0.1:6379> TYPE greet_list
string
```

2.1.2　SETNX

SETNX 命令的基本语法如下。

```
SETNX key value
```

SETNX 是 Set If Not Exists（如果不存在，则 SET）的简写。SETNX 用于将 key 的值设为 String 类型的 value，当 key 不存在时，返回 1；若 key 已经存在，则 SETNX 不执行任何操作，返回 0。

返回值：设置成功，返回 1；设置失败，返回 0。

实例如下。

```
127.0.0.1:6379> EXISTS language            # language 不存在
(integer) 0
127.0.0.1:6379> SETNX language "java"      # language 设置成功
(integer) 1
127.0.0.1:6379> SETNX language "python"    # language 设置失败
(integer) 0
127.0.0.1:6379> GET language               # language 没有被覆盖
"java"
```

第一次设置 language 时没有对应的值，所以 SETNX 修改生效，返回值为 1；第二次设置 language 时已经有了对应的值 java，所以本次修改不生效，返回值为 0。

2.1.3　SETEX

SETEX 命令的基本语法如下。

```
SETEX key seconds value
```

SETEX 用于设置 key 对应的值为 String 类型的 value，并指定此 key 对应的有效期，有效期的过期时间以秒（seconds）为单位。

如果 key 对应的值已经存在，那么 SETEX 将覆盖旧值。这个命令类似于以下两个命令。

```
SET key value              #设置值
EXPIRE key seconds         # 设置过期时间
```

不同之处在于，SETEX 命令是一个原子性操作，设置值和设置过期时间两个操作会在同一时间内完成。该命令经常用在缓存操作中。

返回值：设置成功时返回 OK；当 seconds 参数不合法时，返回一个错误。

实例 1：key 不存在。

```
127.0.0.1:6379> SETEX color 60 red         # 设置 color 的过期时间为 60s
OK
127.0.0.1:6379> GET color                  # 获得 color 值
"red"
127.0.0.1:6379> TTL color                  # 获得 color 剩余的过期时间
(integer) 49
127.0.0.1:6379> GET color
"red"
127.0.0.1:6379> GET color                  # 60s 后，color 值为空
(nil)
```

本例我们添加了一个 color-red 对，并指定它的有效期是 60s。然后使用 TTL 命令查看 color 的过期时间，最后一次调用 color 是在 60s 以后，所以取不到 color 对应的值。

实例 2：key 已经存在，key 对应的值将被覆盖。

```
127.0.0.1:6379> SET color "red"
OK
127.0.0.1:6379> SETEX color 60 "green"
OK
127.0.0.1:6379> GET color
"green"
```

在本例中已经设置了 color 对应的值，可以使用 SETEX 覆盖 color 对应的值。使用 SETEX 设置过期时间 60s，在 60s 内 color 对应的值为 green。

2.1.4　SETRANGE

SETRANGE 命令的基本语法如下。

```
SETRANGE key offset value
```

通过 SETRANGE 用 value 重写 key 所存储的字符串值，从偏移量 offset 开始。不存在的 key 当作空白字符串处理。

返回值：被 SETRANGE 修改之后，字符串的长度。

实例 1：对非空字符串执行 SETRANGE 命令。

例如我们希望将 xpws2006 的 163 邮箱替换为 QQ 邮箱，我们可以这么做。

```
127.0.0.1:6379> SET email "xpws2006@163.com"
OK
127.0.0.1:6379> SETRANGE email 9 "qq.com"
(integer) 18
127.0.0.1:6379> GET email
"xpws2006@qq.com"
```

实例 2：对空字符串/不存在的 key 执行 SETRANGE 命令。

```
127.0.0.1:6379> EXISTS empty_string
(integer) 0
127.0.0.1:6379> SETRANGE empty_string 5 "Redis"   # 对不存在的 key 使用 SETRANGE
(integer) 10
127.0.0.1:6379> GET empty_string   # 空白处被零比特 "\x00" 填充
"\x00\x00\x00\x00\x00Redis"
```

2.1.5　MSET

MSET 命令的基本语法如下。

```
MSET key value [key value ...]
```

通过 MSET 可一次设置多个 key 的值，执行成功返回 OK，表示所有值都被设置了；执行失败返回 0，表示没有任何值被设置。

MSET 是一个原子性操作，所有的 key 都在同一时间内被设置。

返回值：成功返回 OK，失败返回 0。

实例如下。

```
127.0.0.1:6379> MSET name1 "xinping1" name2 "xinping2"
OK
```

```
127.0.0.1:6379> KEYS *                    # 确保指定的两个键值对被插入
1) "name2"
2) "name1"
127.0.0.1:6379> MSET name2 "xinping3"     # MSET 覆盖旧值
OK
127.0.0.1:6379> GET name2
"xinping2"
```

2.1.6 MSETNX

MSETNX 命令的基本语法如下。

```
MSETNX key value [key value ...]
```

MSETNX 用于设置一个或多个 key 的值，执行成功返回 OK，表示所有值都被设置了；执行失败返回 0，表示没有任何值被设置，不会覆盖已经存在的 key。

MSETNX 是原子性的，因此它可以用作设置多个不同的 key，表示不同字段（field）的唯一性逻辑对象（Unique Logic Object），所有字段要么全被设置，要么全不被设置。

返回值：如果所有 key 都成功设置，那么返回 1；如果所有 key 都设置失败（最少有一个 key 已经存在），那么返回 0。

实例 1：对不存在的 key 执行 MSETNX 命令。

```
127.0.0.1:6379> MSETNX key1 "a" key2 "b"
OK
127.0.0.1:6379> GET key1
"a"
127.0.0.1:6379> GET key2
"b"
```

实例 2：对已存在的 key 执行 MSETNX 命令。

```
127.0.0.1:6379> MSET key1 "a" key2 "b"
OK
127.0.0.1:6379> GET key1
"a"
127.0.0.1:6379> GET key2
"b"
127.0.0.1:6379> MSETNX key2 "new_b" key3 "c"   # key2 已经存在，所以操作失败
(integer) 0
127.0.0.1:6379> EXISTS key3                    # 因为命令是原子性的，所以 key3 没有被设置
(integer) 0
127.0.0.1:6379> MGET key1 key2 key3            # key2 没有被修改
1) "a"
2) "b"
3) (nil)
127.0.0.1:6379> GET key3
(nil)
```

2.1.7 APPEND

APPEND 命令的基本语法如下。

```
APPEND key value
```

如果 key 已经存在并且是一个字符串，那么可以通过 APPEND 将 value 追加到 key 关联

的值后面。如果 key 不存在,就简单地将 key 设为 value,就像执行 SET key value 一样。

返回值:追加 value 之后,key 中字符串的长度。

实例 1:对不存在的 key 执行 APPEND 命令。

```
127.0.0.1:6379> EXISTS myphone   # 确保 myphone 不存在
(integer) 0

127.0.0.1:6379> APPEND myphone "huawei"   # 对不存在的 key 执行 APPEND 命令,等同于 SET myphone "huawei"
(integer) 6 # 字符串的长度
```

实例 2:对已存在的 key 执行 APPEND 命令。

```
127.0.0.1:6379> APPEND myphone " p20"
(integer) 10   # 长度从 6 个字符增加到 10 个字符

127.0.0.1:6379> GET myphone   # 查看整个字符串
"Huawei p20"
```

2.1.8　GET

GET 命令的基本语法如下。

```
GET key
```

GET 用于返回 key 所关联的字符串值。如果 key 不存在则返回特殊值 nil。

假如 key 存储的值不是字符串类型,会返回一个错误,因为 GET 只能用于处理字符串值。

返回值:key 的值。如果 key 不存在,返回 nil。

实例 1:获取一个库中已存在的 phone,可以得到它对应的 value。

```
127.0.0.1:6379> SET phone "huawei p20"
OK
127.0.0.1:6379> GET phone
"huawei p20"
```

实例 2:获取一个库中不存在的 phone2,那么它会返回一个 nil,表示没有这个 key-value 对。

```
127.0.0.1:6379> GET phone2
(nil)
```

2.1.9　MGET

MGET 命令的基本语法如下。

```
MGET key [key ...]
```

MGET 用于返回一个或多个 key 的值。如果 key 不存在,那么返回特殊值 nil。因此,该命令永远不会执行失败。

返回值:执行成功则返回一个包含所有 key 的值的列表,执行失败则返回 nil。

实例如下。

```
#用 MSET 一次存储多个值
127.0.0.1:6379> MSET name "xinping"  age 25
OK
127.0.0.1:6379> MGET name age
1) "xinping"
2) "25"
127.0.0.1:6379> EXISTS fake_key
```

```
(integer) 0
# 当 MGET 中有不存在 key 的情况
127.0.0.1:6379> MGET name fake_key
1) "xinping"
2) (nil)
```

2.1.10　GETRANGE

GETRANGE 命令的基本语法如下。

```
GETRANGE key start end
```

GETRANGE 用于获取指定 key 中字符串值的子字符串，子字符串的截取范围由 start 和 end 两个偏移量决定（包括 start 和 end 在内）。负数偏移量表示从字符串的最后开始计数，−1 表示字符串中最后一个字符，−2 表示字符串中倒数第二个字符，其他负数依此类推。

返回值：截取的子字符串。

实例如下。

```
127.0.0.1:6379> SET email "xpws2006@163.com"
OK
127.0.0.1:6379> GET email
"xpws2006@163.com"
127.0.0.1:6379> GETRANGE email 0 7
"xpws2006"
```

GETRANGE email 0 7 截取子字符串的索引是 0～7，包括 0 和 7。

截取子字符串−7～−1，包括−7 和−1。

```
127.0.0.1:6379> GETRANGE email -7 -1
"163.com"
```

截取子字符串从第一个字符到最后一个字符。

```
127.0.0.1:6379> GETRANGE email 0 -1
"xpws2006@163.com"
```

GETRANGE 的取值范围不超过实际字符串长度，超过部分会被忽略。

```
127.0.0.1:6379> GETRANGE email 0 199
"xpws2006@163.com"
```

2.1.11　GETSET

GETSET 命令的基本语法如下。

```
GETSET key value
```

GETSET 用于将 key 的值设为 value，并返回 key 的旧值。

返回值：返回 key 的旧值。当 key 没有旧值时，返回 nil。

实例如下。

```
127.0.0.1:6379> SET name xinping
OK
127.0.0.1:6379> GET name
"xinping"

# name 对应的值被更新，旧值被返回
127.0.0.1:6379> GETSET name xinping_new
"xinping"
```

```
127.0.0.1:6379> GET name
"xinping_new"
```

接下来看一看，如果 key 不存在，那么使用 GETSET 会返回什么值？

```
127.0.0.1:6379> EXISTS name1
(integer) 0
127.0.0.1:6379> GETSET name1 "xinping"
(nil)
127.0.0.1:6379> GET name1
"xinping"
```

因为 name1 之前不存在，没有旧值，所以返回 nil。

GETSET 可以和 INCR 组合使用，实现一个有原子性复位操作功能的计数器（counter）。可以用 GETSET mycount 0 来实现这一目标。

```
127.0.0.1:6379> INCR mycount
(integer) 1

# 一个原子操作内完成 GET mycount 和 GETSET mycount 0
127.0.0.1:6379> GETSET mycount 0
"1"

127.0.0.1:6379> GET mycount
"0"
```

2.1.12　STRLEN

STRLEN 命令的基本语法如下。

```
STRLEN key
```

STRLEN 用于返回 key 所存储的字符串的长度。

返回值：字符串的长度。当 key 不存在时，返回 0。

实例 1：获取 key 存储的字符串 "hello world" 的长度。

```
127.0.0.1:6379> SET key "hello world"
OK
127.0.0.1:6379> STRLEN key
(integer) 11
```

实例 2：当 key 不存在时，它获取的字符串长度为 0。

```
127.0.0.1:6379> STRLEN nonexisting
(integer) 0
```

2.1.13　DECR

DECR 命令的基本语法如下。

```
DECR key
```

DECR 用于将 key 中存储的数值减 1。如果 key 不存在，则以 0 为 key 的初始值，然后执行 DECR 命令，设置 key 对应的值为 –1。

返回值：执行 DECR 命令之后 key 的值。

实例 1：对存在的 key 执行 DECR 命令。

```
127.0.0.1:6379> SET age 23
OK
```

```
127.0.0.1:6379> DECR age
(integer) 22
```
实例 2：对不存在的 key 执行 DECR 命令。
```
127.0.0.1:6379> EXISTS count
(integer) 0
127.0.0.1:6379> DECR count
(integer) -1
```
实例 3：对存在但不是数值的 key 执行 DECR 命令。
```
127.0.0.1:6379> SET name "xinping"
OK
127.0.0.1:6379> DECR company
(error) ERR value is not an integer or out of range
```

2.1.14　DECRBY

DECRBY 命令的基本语法如下。
```
DECRBY key decrement
```
DECRBY 用于将 key 所存储的值减去减量 decrement，也就是指定数值。

如果 key 不存在，则以 0 为 key 的初始值，然后执行 DECRBY 命令。

返回值：减去减量之后 key 的值。

实例 1：对存在的 key 执行 DECRBY 命令。
```
127.0.0.1:6379> SET count 100
OK
127.0.0.1:6379> DECRBY count 20
(integer) 80
```
也可以通过 INCRBY 一个负值来实现同样的效果。
```
127.0.0.1:6379> GET count
"80"
127.0.0.1:6379> INCRBY count -20
(integer) 60
127.0.0.1:6379> GET count
"60"
```
实例 2：对不存在的 key 执行 DECRBY 命令。
```
127.0.0.1:6379> EXISTS pages
(integer) 0
127.0.0.1:6379> DECRBY pages 10
(integer) -10
```

2.1.15　INCR

INCR 命令的基本语法如下。
```
INCR key
```
INCR 用于将 key 中存储的数值增 1。

如果 key 不存在，则以 0 为 key 的初始值，然后执行 INCR 命令，设置 key 为 1。

返回值：执行 INCR 命令之后 key 的值。

实例如下。
```
127.0.0.1:6379> SET age 20
```

```
OK
127.0.0.1:6379> INCR age
(integer) 21
127.0.0.1:6379> GET age
"21"
```

数值在 Redis 中以字符串的形式保存。

2.1.16　INCRBY

INCRBY 命令的基本语法如下。

`INCRBY key increment`

INCRBY 用于将 key 所存储的值加上增量 increment。

如果 key 不存在，则以 0 为 key 的初始值，然后执行 INCRBY 命令。

返回值：加上增量之后 key 的值。

实例 1：key 存在且是数字。

```
127.0.0.1:6379> SET age 21 # 设置 age 为 21
OK
127.0.0.1:6379> INCRBY age 5 # 给 age 加上 5
(integer) 26
127.0.0.1:6379> GET age
"26"
```

实例 2：key 不存在。

```
127.0.0.1:6379> EXISTS counter
(integer) 0
127.0.0.1:6379> INCRBY counter 30
(integer) 30
127.0.0.1:6379> GET counter
"30"
```

实例 3：key 不是数字，那么返回一个错误。

```
127.0.0.1:6379> SET book "how to master redis"
OK
127.0.0.1:6379> INCRBY book 100
(error) ERR value is not an integer or out of range
```

2.2　Hash 类型

Redis 的 Hash 类型是一个 String 类型的域（field）和 value 的映射表，Hash 类型特别适用于存储对象，例如 Username、Password 和 Age 等。

Redis 中的每个 Hash 类型数据都可以存储 $2^{32}-1$ 个 field-value 对。

2.2.1　HSET

HSET 命令的基本语法如下。

`HSET key field value`

HSET 用于将散列表 key 中的 field 的值设置为 value。

返回值：如果散列表 key 中的 field 不存在并且设置成功，则返回 1；如果散列表 key 中的 field 已经存在并且新值覆盖了旧值，则返回 0。

实例如下。

```
127.0.0.1:6379> HSET user name "xinping"        # 创建一个新域
(integer) 1
127.0.0.1:6379> HSET user name "wangwu"         # 覆盖一个旧域
(integer) 0
```

2.2.2 HSETNX

HSETNX 命令的基本语法如下。

`HSETNX key field value`

HSETNX 用于将散列表 key 中的 field 的值设置为 value。如果 key 不存在，那么一个新散列表将被创建并执行 HSETNX 命令，先创建 key。NX 是 Not Exist 的意思。

如果 field 已经存在，则返回 0，该命令无效。

返回值：如果设置 field 成功，则返回 1；如果 field 已经存在，则返回 0。

实例如下。

```
127.0.0.1:6379> HSETNX nosql name "redis"
(integer) 1
127.0.0.1:6379> HSETNX nosql name "redis"       # 命令无效，name 已存在
(integer) 0
```

2.2.3 HMSET

HMSET 命令的基本语法如下。

`HMSET key field value [field value ...]`

HMSET 用于同时将多个 field-value 对设置到散列表 key 中，此命令会覆盖散列表中已存在的 field。

返回值：如果命令执行成功，则返回 OK。

实例 1：将多个 field-value 对设置到散列表 key 中。

```
127.0.0.1:6379> HMSET website taobao "www.taobao.com" jd "www.jd.com"
OK
127.0.0.1:6379> HGET website taobao
"www.taobao.com"
127.0.0.1:6379> HGET website jd
"www.jd.com"
```

实例 2：将 String 类型转为 Hash 类型时，会出现类型转换错误。

```
127.0.0.1:6379> SET user 20
OK
127.0.0.1:6379> HMSET user name wangwu age 21
(error) WRONGTYPE Operation against a key holding the wrong kind of value
```

2.2.4 HGET

HGET 命令的基本语法如下。

`HGET key field`

HGET 用于返回散列表 key 中 field 的值。

返回值：field 的值。当 field 不存在或是 key 不存在时，返回 nil。

实例如下。
```
127.0.0.1:6379> HMSET user name "xinping" age 25
OK
127.0.0.1:6379> HGET user name
"xinping"
127.0.0.1:6379> HGET user age
"25"
127.0.0.1:6379> HGET user address
(nil)
```
由于散列表 key 中没有 address，因此取到的是 nil。

2.2.5 HMGET

HMGET 命令的基本语法如下。
```
HMGET key field [field ...]
```
HMGET 用于返回散列表 key 中一个或多个 field 的值。

返回值：一个或多个给定 field 的值。

实例如下。
```
127.0.0.1:6379> HMSET pet dog "wangwang" cat "miaomiao" # 一次在散列表中保存多个值
OK
127.0.0.1:6379> HMGET pet dog cat fake_pet # 返回值的顺序和传入参数的顺序一样
1) "wangwang"
2) "miaomiao"
3) (nil)
```
由于散列表 key 中没有 fake_pet，因此取到的是 nil。

2.2.6 HGETALL

HGETALL 命令的基本语法如下。
```
HGETALL key
```
HGETALL 用于返回散列表 key 中所有的域和值。

在返回值里，紧跟每个域名（Field Name）之后的是域的值，所以返回值的长度是散列表长度的两倍。

返回值：以列表形式返回散列表 key 的域和值。若 key 不存在，则返回空列表（Empty List）。

实例如下。
```
127.0.0.1:6379> HSET hash_name jd "www.jd.com"
(integer) 1
127.0.0.1:6379> HSET hash_name taobao "www.taobao.com"
(integer) 1
127.0.0.1:6379> HGETALL hash_name
1) "jd"                  # 域
2) "www.jd.com"          # 值
3) "taobao"              # 域
4) "www.taobao.com"      # 值
```

2.2.7 HDEL

HDEL 命令的基本语法如下。
```
HDEL key field [field ...]
```

HDEL 用于删除散列表 key 中的一个或多个 field，不存在的 field 将被忽略。

返回值：被成功删除的 field 的数量。

实例如下。

```
# 设置散列表的测试数据
127.0.0.1:6379> HMSET abbr a "apple" b "banana" c "cat" d "dog"
OK
127.0.0.1:6379> HGETALL abbr
1) "a"
2) "apple"
3) "b"
4) "banana"
5) "c"
6) "cat"
7) "d"
8) "dog"

# 删除单个 field
127.0.0.1:6379> HDEL abbr a
(integer) 1

# 删除不存在的 field
127.0.0.1:6379> HDEL abbr not-exists-field
(integer) 0

# 删除多个 field
127.0.0.1:6379> HDEL abbr b c
(integer) 2

127.0.0.1:6379> HGETALL abbr
1) "d"
2) "dog"
```

2.2.8　HLEN

HLEN 命令的基本语法如下。

```
HLEN key
```

HLEN 用于返回散列表 key 中 field 的数量。

返回值：散列表 key 中 field 的数量。当 key 不存在时，返回 0。

实例如下。

```
127.0.0.1:6379> HSET user name "xinping"
(integer) 1
127.0.0.1:6379> HSET user age 25
(integer) 1
127.0.0.1:6379> HLEN user
(integer) 2
```

2.2.9　HEXISTS

HEXISTS 命令的基本语法如下。

```
HEXISTS key field
```

HEXISTS 用于查看散列表 key 中 field 是否存在。

返回值：查看散列表 key 中，field 如果存在则返回 1，如果不存在则返回 0。

实例如下。

```
127.0.0.1:6379> HEXISTS phone brand
(integer) 0

127.0.0.1:6379> HSET phone brand "xiaomi"
(integer) 1

127.0.0.1:6379> HEXISTS phone brand
(integer) 1
```

2.2.10 HINCRBY

HINCRBY 命令的基本语法如下。

```
HINCRBY key field increment
```

HINCRBY 用于将散列表 key 中的 field 的值加上增量 increment。增量 increment 可以是负数，即对 field 进行减法操作。

返回值：执行 HINCRBY 命令之后，散列表 key 中 field 的值。

实例 1：给指定的 field 加上正数。

```
127.0.0.1:6379> HEXISTS page counter
(integer) 0
127.0.0.1:6379> HINCRBY page counter 20
(integer) 20
127.0.0.1:6379> HGET page counter
"20"
```

实例 2：给指定的 field 加上负数。

```
127.0.0.1:6379> HGET counter page_view
"200"

127.0.0.1:6379> HINCRBY counter page_view -50
(integer) 150

127.0.0.1:6379> HGET counter page_view
"150"
```

实例 3：尝试对字符串值的 field 执行 HINCRBY 命令。

```
127.0.0.1:6379> HSET user name "xinping"          # 对 field 设定一个字符串值
(integer) 1
127.0.0.1:6379> HGET user name
"xinping"
127.0.0.1:6379> HINCRBY user name 1               # 命令执行失败，错误
(error) ERR hash value is not an integer
127.0.0.1:6379> HGET user name                    # 原值不变
"xinping"
```

2.2.11 HKEYS

HKEYS 命令的基本语法如下。

```
HKEYS key
```
HKEYS 用于返回散列表 key 中的所有域。

返回值：一个列表，该列表包含散列表 key 中的所有域。当 key 不存在时，返回一个空列表。

实例：返回散列表 key 中的所有域。
```
127.0.0.1:6379> HMSET website jd "www.jd.com"  taobao "www.taobao.com"
OK
127.0.0.1:6379> HKEYS website
1) "jd"
2) "taobao"
```
散列表 website 中有两个域。

2.2.12　HVALS

HVALS 命令的基本语法如下。
```
HVALS key
```
HVALS 用于返回散列表 key 中的所有值。

返回值：当散列表 key 存在时，返回一个列表，该列表包含散列表 key 中的所有值；当散列表 key 不存在时，返回一个空列表。

实例：返回散列表 key 中的所有值。
```
127.0.0.1:6379> HMSET website jd "www.jd.com"  taobao "www.taobao.com"
OK
127.0.0.1:6379> HVALS website
1) "www.jd.com"
2) "www.taobao.com"
```

2.3　List 类型

在 Redis 中，List 类型是按照元素的插入顺序排序的字符串列表。在插入时，如果 key 并不存在，Redis 将为该 key 创建一个新的列表。List 类型中可以包含的最大元素数量是 4 294 967 295。

2.3.1　LPUSH

LPUSH 命令的基本语法如下。
```
LPUSH key value [value ...]
```
LPUSH 用于将一个或多个 value 插入列表 key 的表头，可以作为栈，特点是先进后出。

返回值：执行 LPUSH 命令后，列表 key 的长度。

实例：对空列表执行 LPUSH 命令。
```
127.0.0.1:6379> DEL mykey                       # 删除一个 key 为 mykey 的列表
(integer) 0
127.0.0.1:6379> LPUSH mykey a
(integer) 1
127.0.0.1:6379> LPUSH mykey b
(integer) 2
127.0.0.1:6379> LPUSH mykey c
(integer) 3
```

```
127.0.0.1:6379> LPUSH mykey d
(integer) 4
```
使用 LPUSH 将 3 个值插入名为 mykey 的列表当中,也可以一次插入多个值到列表,效果是一样的。
```
127.0.0.1:6379> DEL mykey
(integer) 1
127.0.0.1:6379> LPUSH mykey a b c d
(integer) 4
```

2.3.2　LPUSHX

LPUSHX 命令的基本语法如下。

```
LPUSHX key value
```

LPUSHX 用于将 value 插入 key,key 存在并且是一个列表。

和 LPUSH 命令相反,当 key 不存在时,LPUSHX 什么也不做。

返回值:执行 LPUSHX 命令之后,列表 key 的长度。

实例 1:对空列表执行 LPUSHX 命令。

```
127.0.0.1:6379> LLEN mylist          # mylist 是一个空列表
(integer) 0
127.0.0.1:6379> LPUSHX mylist 1      # 尝试执行 LPUSHX 命令,失败,因为列表为空
(integer) 0
```

实例 2:对非空列表执行 LPUSHX 命令。

```
127.0.0.1:6379> LPUSH mylist 1       # 先用 LPUSH 创建一个有一个元素的列表
(integer) 1
127.0.0.1:6379> LPUSHX mylist 2      # 这次 LPUSHX 命令执行成功
(integer) 2
127.0.0.1:6379> LRANGE mylist 0 -1
1) "2"
2) "1"
```

2.3.3　RPUSH

RPUSH 命令的基本语法如下。

```
RPUSH key value [value ...]
```

RPUSH 用于将一个或多个 value 插入列表 key 的表尾,可以作为队列,特点是先进先出。

返回值:执行 RPUSH 命令后,列表 key 的长度。

实例如下。

```
# 删除已经存在的 key(mylist)
127.0.0.1:6379> DEL mylist
(integer) 1

# 添加单个元素
127.0.0.1:6379> RPUSH  mylist 1
(integer) 1

# 添加重复元素
127.0.0.1:6379> RPUSH mylist 2
(integer) 2
```

```
# 列表允许重复元素
127.0.0.1:6379> LRANGE mylist 0 -1
1) "1"
2) "2"

# 添加多个元素
127.0.0.1:6379> LPUSH mylist a b c
(integer) 5

127.0.0.1:6379> LRANGE mylist 0 -1
1) "c"
2) "b"
3) "a"
4) "1"
5) "2"
```

2.3.4 RPUSHX

RPUSHX 命令的基本语法如下。

```
RPUSHX key value
```

RPUSHX 用于将 value 插入列表 key 的表尾，并且列表 key 存在。

和 RPUSH 相反，当 key 不存在时，RPUSHX 什么也不做。

返回值：执行 RPUSHX 命令之后，列表 key 的长度。

实例 1：key 不存在。

```
127.0.0.1:6379> LLEN greet
(integer) 0

127.0.0.1:6379> RPUSHX greet "hello"   # 对不存在的 key 执行 RPUSHX 命令，失败
(integer) 0
```

实例 2：key 存在且是一个非空列表。

```
127.0.0.1:6379> RPUSH greet "hi"   # 先用 RPUSH 插入一个元素
(integer) 1

127.0.0.1:6379> RPUSHX greet "hello"   # greet 现在是一个列表类型，执行 RPUSHX 命令成功
(integer) 2

127.0.0.1:6379> LRANGE greet 0 -1
1) "hi"
2) "hello"
```

2.3.5 LPOP

LPOP 命令的基本语法如下。

```
LPOP key
```

LPOP 用于从列表 key 的头部删除元素，并返回删除元素。

返回值：列表 key 的头元素。当 key 不存在时，返回 nil。

实例如下。

```
127.0.0.1:6379> LLEN course
```

```
(integer) 0
127.0.0.1:6379> RPUSH course java
(integer) 1
127.0.0.1:6379> RPUSH course python
(integer) 2
127.0.0.1:6379> LRANGE course 0 -1
1) "java"
2) "python"
# 删除列表的头元素
127.0.0.1:6379> LPOP course
"java"
127.0.0.1:6379> LRANGE course 0 -1
1) "python"
```

2.3.6　RPOP

RPOP 命令的基本语法如下。

```
RPOP key
```

RPOP 用于从列表 key 的尾部删除元素，并返回删除元素。

返回值：列表 key 的尾元素。当 key 不存在时，返回 nil。

实例如下。

```
127.0.0.1:6379> LLEN mylist
(integer) 0
127.0.0.1:6379> RPUSH mylist "one"
(integer) 1
127.0.0.1:6379> RPUSH mylist "two"
(integer) 2
127.0.0.1:6379> RPUSH mylist "three"
(integer) 3
# 返回被删除的元素
127.0.0.1:6379> RPOP mylist
"three"
# 列表剩下的元素
127.0.0.1:6379> LRANGE mylist 0 -1
1) "one"
2) "two"
```

2.3.7　LLEN

LLEN 命令的基本语法如下。

```
LLEN key
```

LLEN 用于返回列表 key 的长度。

如果 key 不存在，则 key 被解释为一个空列表，返回 0。如果 key 不是 List 类型，返回一个错误。

返回值：列表 key 的长度。

实例：返回非空列表的长度。

```
127.0.0.1:6379> LPUSH course "java"
(integer) 1
127.0.0.1:6379> LPUSH course "python"
```

```
(integer) 2
127.0.0.1:6379> LLEN course
(integer) 2
```

2.3.8　LREM

LREM 命令的基本语法如下。

```
LREM key count value
```

LREM 用于从列表 key 中删除 count 个和 value 相等的元素。

count 的值可以是以下几种。

- count > 0：从列表的表头开始向表尾遍历，删除与 value 相等的元素，数量为 count。
- count < 0：从列表的表尾开始向表头遍历，删除与 value 相等的元素，数量为 count。
- count = 0：删除列表中所有与 value 相等的元素。

返回值：被删除元素的数量。

因为不存在的 key 被视作空列表，所以当 key 不存在时，LREM 总是返回 0。

实例如下。

```
# 先创建一个列表，元素排列如下
# morning hello morning hello morning

127.0.0.1:6379> LPUSH greet "morning"
(integer) 1
127.0.0.1:6379> LPUSH greet "hello"
(integer) 2
127.0.0.1:6379> LPUSH greet "morning"
(integer) 3
127.0.0.1:6379> LPUSH greet "hello"
(integer) 4
127.0.0.1:6379> LPUSH greet "morning"
(integer) 5

# 查看所有元素
127.0.0.1:6379> LRANGE greet 0 4
1) "morning"
2) "hello"
3) "morning"
4) "hello"
5) "morning"

# 删除从表头到表尾最先发现的两个 morning
127.0.0.1:6379> LREM greet 2 morning
(integer) 2          # 两个元素被删除

# 还剩 3 个元素
127.0.0.1:6379> LLEN greet    (integer) 3

127.0.0.1:6379> LRANGE greet 0 2
1) "hello"
2) "hello"
3) "morning"
```

count < 0 时，按从表尾到表头的顺序删除元素，具体如下。

```
# 删除从表尾到表头的元素，第一个是 morning
127.0.0.1:6379> LREM greet -1 morning
(integer) 1

127.0.0.1:6379> LLEN greet
(integer) 2
```

count = 0 时，删除全部元素，具体如下。

```
127.0.0.1:6379> LRANGE greet 0 1
1) "hello"
2) "hello"

# 删除列表中所有 hello
127.0.0.1:6379> LREM greet 0 hello
(integer) 2        # 两个 hello 被删除

127.0.0.1:6379> LLEN greet
(integer) 0
```

2.3.9 LSET

LSET 命令的基本语法如下。

```
LSET key index value
```

LSET 用于设置列表 key 中指定索引的元素值，索引从 0 开始计数。

返回值：执行成功则返回 OK，否则返回错误信息。

实例 1：对空列表（key 不存在）执行 LSET 命令。

```
127.0.0.1:6379> EXISTS list
(integer) 0
127.0.0.1:6379> LSET list 0 one
(error) ERR no such key
```

实例 2：对非空列表执行 LSET 命令。

```
127.0.0.1:6379> LPUSH list "one"
(integer) 1
127.0.0.1:6379> LPUSH list "two"
(integer) 2
127.0.0.1:6379> LRANGE list 0 -1
1) "two"
2) "one"
127.0.0.1:6379> LSET list 0 "three"
OK
127.0.0.1:6379> LRANGE list 0 -1
1) "three"
2) "one"
```

实例 3：索引超出范围。

```
127.0.0.1:6379> LLEN list  # 列表长度为 2
(integer) 2
127.0.0.1:6379> LSET 3 "three"
(error) ERR wrong number of arguments for 'lset' command
```

2.3.10　LTRIM

LTRIM 命令的基本语法如下。

```
LTRIM key start stop
```

LTRIM 用于对列表 key 进行修剪，让列表 key 只保留指定区间内的元素，不在列表 key 指定区间之内的元素都将被删除。举个例子，执行命令 LTRIM list 0 2，表示只保留列表 list 的前 3 个元素，其余元素全部被删除。

当 key 不是 List 类型时，返回一个错误。

返回值：命令执行成功时，返回 OK。

实例 1：一般情况下的索引。

```
# 建立一个 4 个元素的列表
127.0.0.1:6379> RPUSH list2 1
(integer) 1
127.0.0.1:6379> RPUSH list2 2
(integer) 2
127.0.0.1:6379> RPUSH list2 3
(integer) 3
127.0.0.1:6379> RPUSH list2 4
(integer) 4
# 删除索引为 0 的元素
127.0.0.1:6379> LTRIM list2 1 -1
OK
# "1" 被删除
127.0.0.1:6379> LRANGE list2 0 -1
1) "2"
2) "3"
3) "4"
```

实例 2：stop 比元素的最大索引要大。

```
127.0.0.1:6379> DEL list2
(integer) 1
127.0.0.1:6379> RPUSH list2 1
(integer) 1
127.0.0.1:6379> RPUSH list2 2
(integer) 2
127.0.0.1:6379> RPUSH list2 3
(integer) 3
127.0.0.1:6379> RPUSH list2 4
(integer) 4
127.0.0.1:6379> LTRIM list2 1 100
OK
127.0.0.1:6379> LRANGE list2 0 -1
1) "2"
2) "3"
3) "4"
```

实例 3：start 和 stop 都比最大索引要大，且 start < stop。

```
# 整个列表被清空，等同于 DEL list2
127.0.0.1:6379> LTRIM list2 100 200
```

```
OK
127.0.0.1:6379> LRANGE list2 0 -1
(empty list or set)
```
实例 4：start > stop。
```
# 新建一个列表 list2
127.0.0.1:6379> DEL list2
(integer) 1
127.0.0.1:6379> RPUSH list2 1
(integer) 1
127.0.0.1:6379> RPUSH list2 2
(integer) 2
127.0.0.1:6379> RPUSH list2 3
(integer) 3
127.0.0.1:6379> RPUSH list2 4
(integer) 4

# 列表 list2 同样被清空
127.0.0.1:6379> LTRIM list2 1000 4
OK
127.0.0.1:6379> LRANGE list2 0 -1
(empty list or set)

127.0.0.1:6379> LRANGE alpha 0 -1 # 再新建一个列表
1) "h"
2) "u"
3) "a"
4) "n"
5) "g"
6) "z"
```

2.3.11 LINDEX

LINDEX 命令的基本语法如下。
```
LINDEX key index
```
LINDEX 用于返回名称为 key 的列表中 index 位置的元素。如果 key 不是 List 类型，返回一个错误。

返回值：列表 key 中索引为 index 的元素。

实例如下。
```
127.0.0.1:6379> RPUSH list3 "a"
(integer) 1
127.0.0.1:6379> RPUSH list3 "b"
(integer) 2
127.0.0.1:6379> LRANGE list3 0 -1
1) "a"
2) "b"
127.0.0.1:6379> LINDEX list3 0
"a"
127.0.0.1:6379> LINDEX list3 -1
"b"
# index 不在列表 list3 的区间范围内会返回 nil
```

```
127.0.0.1:6379> LINDEX list3 3
(nil)
```

2.3.12 LINSERT

LINSERT 命令的基本语法如下。

```
LINSERT key BEFORE|AFTER pivot value
```

LINSERT 用于将 value 插入列表 key 当中,位于 pivot 之前或之后。

当 pivot 不存在于列表 key 时,不执行任何操作。如果 key 不是 List 类型,返回一个错误。

返回值:如果执行 LINSERT 命令成功,则返回执行之后的列表长度;如果没有找到 pivot,则返回-1;如果 key 不存在或为空列表,则返回 0。

实例如下。

```
127.0.0.1:6379> RPUSH mylist "Hello"
(integer) 1
127.0.0.1:6379> RPUSH mylist "World"
(integer) 2

127.0.0.1:6379> LINSERT mylist BEFORE "World" "There"
(integer) 3

127.0.0.1:6379> LRANGE mylist 0 -1
1) "Hello"
2) "There"
3) "World"

# 对一个非空列表插入,查找一个不存在的 pivot
127.0.0.1:6379> LINSERT mylist BEFORE "go" "let's"
(integer) -1    # 失败

127.0.0.1:6379> EXISTS fake_list   # 对一个空列表执行 LINSERT 命令
(integer) 0

127.0.0.1:6379> LINSERT fake_list BEFORE "none" "a"
(integer) 0 # 失败
```

2.3.13 RPOPLPUSH

RPOPLPUSH 命令的基本语法如下。

```
RPOPLPUSH source destination
```

RPOPLPUSH 用于将元素从第一个列表的表尾移动到第二个列表的表头,并返回被移除的元素。整个操作是原子性的,如果第一个列表是空或者不存在则返回 nil。

举个例子,有两个列表 source 和 destination,列表 source 有元素 a、b、c,列表 destination 有元素 x、y、z,执行 RPOPLPUSH source destination 之后,列表 source 包含元素 a、b,列表 destination 包含元素 c、x、y、z,并且元素 c 被返回。

如果列表 source 不存在,则返回 nil,并且不执行其他操作。

如果列表 source 和列表 destination 相同,则列表 source 中的表尾元素被移动到表头,并返回该元素。

返回值：被移除的元素。

实例如下。

```
# 生成列表
127.0.0.1:6379> DEL list1
(integer) 0
127.0.0.1:6379> DEL list2
(integer) 0
127.0.0.1:6379> RPUSH list1 a
(integer) 1
127.0.0.1:6379> RPUSH list1 b
(integer) 2
127.0.0.1:6379> RPUSH list1 c
(integer) 3
127.0.0.1:6379> RPUSH list1 d
(integer) 4
# 查看所有元素
127.0.0.1:6379> LRANGE list1 0 -1
1) "a"
2) "b"
3) "c"
4) "d"
# 情况 1：列表 source 和列表 destination 不同
127.0.0.1:6379> RPOPLPUSH list1 list2
"d"
127.0.0.1:6379> LRANGE list1 0 -1
1) "a"
2) "b"
3) "c"
127.0.0.1:6379> LRANGE list2 0 -1
1) "d"
# 再执行一次 RPOPLPUSH 操作，把列表 List1 的尾部元素、添加到列表 List2 的头部
127.0.0.1:6379> RPOPLPUSH list1 list2
"c"
127.0.0.1:6379> LRANGE list1 0 -1
1) "a"
2) "b"
127.0.0.1:6379> LRANGE list2 0 -1
1) "c"
2) "d"

# 情况 2：列表 source 和列表 destination 相同
127.0.0.1:6379> RPOPLPUSH list1 list1
"b"
# 把列表 list1 原来的表尾元素 "c" 放到表头
127.0.0.1:6379> LRANGE list1 0 -1
1) "c"
2) "d"
```

2.4　Set 类型

Redis 的 Set 类型是 String 类型的无序集合。集合中的元素是唯一的，不能出现重复的元素。

2.4.1 SADD

SADD 命令的基本语法如下。

```
SADD key member [member ...]
```

SADD 用于将一个或多个 member 加入集合 key 当中。

假如 key 不存在，则创建一个只包含 member 的集合。当 key 不是 set 类型时，返回一个错误。

返回值：被添加到集合 key 中的新元素的数量。

实例如下。

```
# 添加单个元素
127.0.0.1:6379> SADD letter a
(integer) 1

# 添加重复元素
127.0.0.1:6379> SADD letter a
(integer) 0

# 添加多个元素
127.0.0.1:6379> SADD letter b c
(integer) 2

# 查看集合
127.0.0.1:6379> SMEMBERS letter
1) "c"
2) "b"
3) "a"
```

本例中，我们向集合 letter 中添加 3 个元素，但是重复元素 a 没有添加成功，最后用 SMEMBERS 查看集合的所有元素。

2.4.2 SREM

SREM 命令的基本语法如下。

```
SREM key member [member ...]
```

SREM 用于删除集合 key 中的一个或多个 member，如果 member 不存在则会被忽略。

当 key 不是 Set 类型时，返回一个错误。

返回值：执行 SREM 命令成功后，返回集合 key 中被成功删除的元素数量。

实例如下。

```
# 添加测试元素
127.0.0.1:6379> SADD myset "one"
(integer) 1
127.0.0.1:6379> SADD myset "two"
(integer) 1
127.0.0.1:6379> SADD myset "three"
(integer) 1
127.0.0.1:6379>
127.0.0.1:6379> SMEMBERS myset
1) "three"
2) "two"
```

```
3) "one"

# 删除单个元素
127.0.0.1:6379> SREM myset "one"
(integer) 1

# 删除不存在的元素
127.0.0.1:6379> SREM myset "none"
(integer) 0

# 删除多个元素
127.0.0.1:6379> SREM myset three two
(integer) 2
127.0.0.1:6379> SMEMBERS myset
(empty list or set)
```
本例中，我们先向集合 myset 中添加了 3 个元素，再执行 SREM 命令来删除 one 和 none。由于集合 myset 中有元素 one，因此 one 被删除，而集合 myset 中没有 none，所以 SREM 命令执行失败。

2.4.3 SMEMBERS

SMEMBERS 命令的基本语法如下。

```
SMEMBERS key
```

SMEMBERS 的返回集合 key 中的所有元素。

返回值：集合 key 中的所有元素。

实例 1：空集合。

```
# 不存在的 key 视为空集合
127.0.0.1:6379> EXISTS not_exists_key
(integer) 0

127.0.0.1:6379> SMEMBERS not_exists_key
(empty list or set)
```

实例 2：非空集合。

```
127.0.0.1:6379> SADD programming_language python
(integer) 1

127.0.0.1:6379> SADD programming_language ruby
(integer) 1

127.0.0.1:6379> SADD programming_language c
(integer) 1

127.0.0.1:6379> SMEMBERS programming_language
1) "c"
2) "ruby"
3) "python"
```

2.4.4 SCARD

SCARD 命令的基本语法如下。

```
SCARD key
```
SCARD 返回集合 key 中元素的数量。

返回值：集合 key 中元素的数量。当 key 不存在时，返回 0。

实例如下。

```
127.0.0.1:6379> SADD myset2 a b c
(integer) 3
127.0.0.1:6379> SCARD myset2
(integer) 3
127.0.0.1:6379> SMEMBERS myset2
1) "c"
2) "b"
3) "a"
```

2.4.5　SMOVE

SMOVE 命令的基本语法如下。

```
SMOVE source destination member
```

SMOVE 用于将 member 从集合 source 移动到集合 destination，也就是从第一个集合中删除 member 并添加到第二个对应集合中。

SMOVE 命令是原子性操作。如果集合 source 不存在，则 SMOVE 不执行任何操作，仅返回 0。否则 member 从集合 source 中被删除，并添加到集合 destination 中。

当集合 destination 已经包含 member 时，SMOVE 只是简单地将集合 source 中的 member 删除。

当 source 或 destination 不是 Set 类型时，返回一个错误。

返回值：如果 member 被成功删除，那么返回 1；如果 member 不是集合 source 的元素，并且没有任何对集合 destination 的操作，那么返回 0。

实例如下。

```
127.0.0.1:6379> DEL myset2
(integer) 1
127.0.0.1:6379> SADD myset2 a b c
(integer) 3
127.0.0.1:6379> SMEMBERS myset2
1) "c"
2) "b"
3) "a"
127.0.0.1:6379> SMOVE myset2 myset3 a
(integer) 1
127.0.0.1:6379> SMEMBERS myset2
1) "c"
2) "b"
127.0.0.1:6379> SMEMBERS myset3
1) "a"
```

通过本例可以看到，集合 myset2 中的元素 a 被移到集合 myset3 中了。

2.4.6　SPOP

SPOP 命令的基本语法如下。

```
SPOP key
```
SPOP 用于随机返回并删除名称为 key 的集合中的一个元素。

返回值：被删除的随机元素。当 key 不存在或 key 是空集时，返回 nil。

实例如下。

```
127.0.0.1:6379> DEL myset3
(integer) 1
127.0.0.1:6379> SADD myset3 "one"
(integer) 1
127.0.0.1:6379> SADD myset3 "two"
(integer) 1
127.0.0.1:6379> SADD myset3 "three"
(integer) 1
127.0.0.1:6379> SPOP myset3
"one"
127.0.0.1:6379> SMEMBERS myset3
1) "three"
2) "two"
```

本例中，我们向集合 myset3 中添加了 3 个元素后，再执行 SPOP 命令来随机删除一个元素，可以看到元素 one 被删除了。

2.4.7　SRANDMEMBER

SRANDMEMBER 命令的基本语法如下。

```
SRANDMEMBER key
```

SRANDMEMBER 用于随机返回名称为 key 的集合中的一个元素，但是不删除元素。

返回值：被选中的随机元素。当 key 不存在或 key 是空集时，返回 nil。

实例如下。

```
127.0.0.1:6379> SADD myset4 "a"
(integer) 1
127.0.0.1:6379> SADD myset4 "b"
(integer) 1
127.0.0.1:6379> SADD myset4 "c"
(integer) 1
127.0.0.1:6379> SADD myset4 "d"
(integer) 1
127.0.0.1:6379> SMEMBERS myset4
1) "d"
2) "c"
3) "b"
4) "a"

127.0.0.1:6379> SRANDMEMBER myset4
"a"
127.0.0.1:6379> SRANDMEMBER myset4
"d"

127.0.0.1:6379> SMEMBERS myset4
1) "d"
```

2) "c"
3) "b"
4) "a"

2.4.8 SINTER

SINTER 命令的基本语法如下。

```
SINTER key [key ...]
```

SINTER 用于返回集合 key 中的交集。

返回值：交集元素的列表。

实例如下。

```
127.0.0.1:6379> SADD myset4 "a"
(integer) 1
127.0.0.1:6379> SADD myset4 "b"
(integer) 1
127.0.0.1:6379> SADD myset5 "b"
(integer) 1
127.0.0.1:6379> SADD myset5 "c"
(integer) 1
127.0.0.1:6379> SMEMBERS myset4
1) "b"
2) "a"
127.0.0.1:6379> SMEMBERS myset5
1) "c"
2) "b"
127.0.0.1:6379> SINTER myset4 myset5
1) "b"
```

通过本例的结果可以看出，集合 myset4 和集合 myset5 的交集元素 b 被找出来了。

2.4.9 SINTERSTORE

SINTERSTORE 命令的基本语法如下。

```
SINTERSTORE destination key [key ...]
```

此命令等同于 SINTER，但它将结果保存到集合 destination，而不是简单地返回结果。

如果集合 destination 已经存在，则将其覆盖。

返回值：交集中的元素数量。

实例如下。

```
127.0.0.1:6379> SADD myset6 "a"
(integer) 1
127.0.0.1:6379> SADD myset6 "b"
(integer) 1
127.0.0.1:6379> SADD myset7 "b"
(integer) 1
127.0.0.1:6379> SADD myset7 "c"
(integer) 1
127.0.0.1:6379> SMEMBERS myset6
1) "b"
2) "a"
127.0.0.1:6379> SMEMBERS myset7
```

```
1) "c"
2) "b"
127.0.0.1:6379> SINTERSTORE myset8 myset6 myset7
(integer) 1
127.0.0.1:6379> SMEMBERS myset8
1) "b"
```

通过本例的结果我们可以看出，集合 myset6 和集合 myset7 的交集被保存到集合 myset8 中了。

2.4.10 SUNION

SUNION 命令的基本语法如下。

```
SUNION key [key ...]
```

SUNION 用于返回所有集合 key 的并集。不存在的 key 被视为空集。

返回值：并集元素的列表。

实例如下。

```
127.0.0.1:6379> DEL myset1
(integer) 0
127.0.0.1:6379> DEL myset2
(integer) 1
127.0.0.1:6379> SADD myset1 a
(integer) 1
127.0.0.1:6379> SADD myset1 b
(integer) 1
127.0.0.1:6379> SADD myset2 b
(integer) 1
127.0.0.1:6379> SADD myset2 c
(integer) 1
127.0.0.1:6379> SUNION myset1 myset2
1) "c"
2) "b"
3) "a"
```

通过本例的结果可以看出，集合 myset1 和集合 myset2 的并集被找出来了。

2.4.11 SUNIONSTORE

SUNIONSTORE 命令的基本语法如下。

```
SUNIONSTORE destination key [key ...]
```

此命令等同于 SUNION，但它将结果保存到集合 destination，而不是简单地返回结果。

如果集合 destination 已经存在，则将其覆盖。集合 destination 可以是集合 key 本身。

返回值：并集中的元素数量。

实例如下。

```
127.0.0.1:6379> DEL myset1
(integer) 1
127.0.0.1:6379> DEL myset2
(integer) 1
127.0.0.1:6379> DEl myset3
(integer) 0
```

```
127.0.0.1:6379> SADD myset1 a
(integer) 1
127.0.0.1:6379> SADD myset1 b
(integer) 1
127.0.0.1:6379> SADD myset2 b
(integer) 1
127.0.0.1:6379> SADD myset2 c
(integer) 1
127.0.0.1:6379> SUNIONSTORE myset3 myset1 myset2
(integer) 3
127.0.0.1:6379> SMEMBERS myset3
1) "c"
2) "b"
3) "a"
```
通过本例的结果可以看出，集合 myset1 和集合 myset2 的并集被保存到集合 myset3 中了。

2.4.12 SDIFF

SDIFF 命令的基本语法如下。

```
SDIFF key [key ...]
```

SDIFF 用于返回集合 key 的差集。不存在的 key 被视为空集。

返回值：差集元素的列表。

实例如下。

```
127.0.0.1:6379> SMEMBERS myset1
1) "b"
2) "a"
127.0.0.1:6379> SMEMBERS myset2
1) "c"
2) "b"
127.0.0.1:6379> SDIFF myset1 myset2
1) "a"
```

从本例中，我们可以看到集合 myset1 和集合 myset2 的差集元素是 a。我们也可以将集合 myset1 和集合 myset2 换个顺序看一下结果。

```
127.0.0.1:6379> SDIFF myset2 myset1
1) "c"
```

从这个结果可以看出，集合 myset2 与集合 myset1 的差集元素是 c。

2.4.13 SDIFFSTORE

SDIFFSTORE 命令的基本语法如下。

```
SDIFFSTORE destination key [key ...]
```

此命令等同于 SDIFF，但它将结果保存到集合 destination，而不是简单地返回结果。

如果集合 destination 已经存在，则将其覆盖。集合 destination 可以是集合 key 本身。

返回值：差集中的元素数量。

实例如下。

```
127.0.0.1:6379> DEL myset1
(integer) 1
127.0.0.1:6379> DEL myset2
```

```
(integer) 1
127.0.0.1:6379> DEl myset3
(integer) 1
127.0.0.1:6379> SADD myset1 a
(integer) 1
127.0.0.1:6379> SADD myset1 b
(integer) 1
127.0.0.1:6379> SADD myset2 b
(integer) 1
127.0.0.1:6379> SADD myset2 c
(integer) 1
127.0.0.1:6379> SMEMBERS myset1
1) "b"
2) "a"
127.0.0.1:6379> SMEMBERS myset2
1) "c"
2) "b"
127.0.0.1:6379> SDIFFSTORE myset3 myset1 myset2
(integer) 1
127.0.0.1:6379> SMEMBERS myset3
1) "a"
```

2.5　Sorted Set 类型

Sorted Set 类型是 Set 类型的一个加强版本，它在 Set 类型的基础上增加了一个顺序属性。这一属性在添加、修改元素的时候可以指定，每次指定后有序集合会自动按新的值调整顺序。

有序集合中的元素是唯一的，但分数（Score）却可以重复。

2.5.1　ZADD

ZADD 命令的基本语法如下。

```
ZADD key score member [[score member] [score member] ...]
```

ZADD 用于将一个或多个 member 和 score 加入有序集合 key 当中。

返回值：被成功添加的新元素的数量。

实例如下。

```
# 添加单个元素
127.0.0.1:6379> ZADD myzset1 1 "one"
(integer) 1

# 添加多个元素
127.0.0.1:6379> ZADD myzset1 2 "two" 3 "three"
(integer) 2

# 显示有序集合 myzset1
127.0.0.1:6379> ZRANGE myzset1 0 -1 WITHSCORES
1) "one"
2) "1"
3) "two"
```

4) "2"
5) "three"
6) "3"

使用 Redis 的内存可视化工具 Redis Desktop Manager 查看有序集合 myzset1 在 Redis 中的存储结构，如图 2-1 所示。

```
# 添加已存在元素，但是改变 score
127.0.0.1:6379> ZADD myzset1 6 "one"
(integer) 0

127.0.0.1:6379> ZRANGE myzset1 0 -1 WITHSCORES
1) "two"
2) "2"
3) "three"
4) "3"
5) "one"
6) "6"
```

在本例中，我们向有序集合 myzset1 中添加了元素 one、two 和 three，并且元素 one 被设置了两次，那么将以最后一次的设置为准。最后我们将所有元素都显示出来，注意观察元素的 score，如图 2-2 所示。

图 2-1　有序集合 myzset1 在 Redis 中的存储结构　　图 2-2　有序集合 myzset1 重新设置了元素 one

2.5.2　ZREM

ZREM 命令的基本语法如下。

`ZREM key member [member ...]`

ZREM 用于删除有序集合 key 中的一个或多个 member，不存在的 member 将被忽略。

返回值：有序集合 key 中被成功删除的元素数量。

实例如下。

```
# 生成有序集合测试数据
127.0.0.1:6379> ZADD myzset2 1 "one"
(integer) 1
127.0.0.1:6379> ZADD myzset2 2 "two"
(integer) 1
127.0.0.1:6379> ZADD myzset2 3 "three"
(integer) 1
127.0.0.1:6379> ZADD myzset2 4 "four"
(integer) 1
127.0.0.1:6379> ZRANGE myzset2 0 -1 WITHSCORES
1) "one"
2) "1"
3) "two"
```

```
4) "2"
5) "three"
6) "3"
7) "four"
8) "4"

# 删除单个元素
127.0.0.1:6379> ZREM myzset2 "two"
(integer) 1
127.0.0.1:6379> ZRANGE myzset2 0 -1
1) "one"
2) "three"
3) "four"

# 删除多个元素
127.0.0.1:6379> ZREM myzset2 one three
(integer) 2
127.0.0.1:6379> ZRANGE myzset2 0 -1
1) "four"

# 删除不存在的元素
127.0.0.1:6379> ZREM myzset2 "five"
(integer) 0
```

2.5.3 ZCARD

ZCARD 命令的基本语法如下。

```
ZCARD key
```

ZCARD 用于返回有序集合 key 中的元素个数。

返回值：当有序集合 key 存在时，返回有序集合 key 的元素个数；当有序集合 key 不存在时，返回 0。

实例如下。

```
# 添加一个元素
127.0.0.1:6379> ZADD salary 5000 wangwu
(integer) 1

# 再添加一个元素
127.0.0.1:6379> ZADD salary 6000 lisi
(integer) 1

127.0.0.1:6379> ZCARD salary
(integer) 2

# 对不存在的 key 执行 ZCARD 命令
127.0.0.1:6379> EXISTS non_exists_key
(integer) 0
127.0.0.1:6379> ZCARD non_exists_key
(integer) 0
```

从本例可以看出，有序集合 salary 的元素个数是 2。

2.5.4 ZCOUNT

ZCOUNT 命令的基本语法如下。

```
ZCOUNT key min max
```

ZCOUNT 用于返回有序集合 key 中 score 值在 min 和 max 之间（默认包括 score 值等于 min 或 max）的元素数量，也就是返回有序集合 key 中 score 值在给定区间的元素数量。

返回值：score 值在 min 和 max 之间的元素数量。

实例如下。

```
# 添加有序集合的元素
127.0.0.1:6379> DEL salary
(integer) 1
127.0.0.1:6379> ZADD salary 3000 wangwu
(integer) 1
127.0.0.1:6379> ZADD salary 4000 lisi
(integer) 1
127.0.0.1:6379> ZADD salary 5000 zhangsan
(integer) 1

# 显示所有元素及其 score 值
127.0.0.1:6379> ZRANGE salary 0 -1 WITHSCORES
1) "wangwu"
2) "3000"
3) "lisi"
4) "4000"
5) "zhangsan"
6) "5000"

# 计算 score 值为 3000~5000 的元素数量
127.0.0.1:6379> ZCOUNT salary 3000 5000
(integer) 3

# 计算 score 值为 4000～5000 元素数量
127.0.0.1:6379> ZCOUNT salary 4000 5000
(integer) 2
```

2.5.5 ZSCORE

ZSCORE 命令的基本语法如下。

```
ZSCORE key member
```

ZSCORE 用于返回有序集合 key 中 member 的 score 值。如果 member 不是有序集合 key 的元素，或有序集合 key 不存在，则返回 nil。

返回值：member 的 score 值，以字符串形式表示。

实例如下。

```
# 显示有序集合中所有元素及其 score 值
127.0.0.1:6379> ZRANGE salary 0 -1 WITHSCORES
1) "wangwu"
2) "3000"
3) "lisi"
```

```
4) "4000"
5) "zhangsan"
6) "5000"

# 注意 score 值以字符串形式表示
127.0.0.1:6379> ZSCORE salary wangwu
"3000"
```

在本例中,我们成功地获取了 wangwu 的 score 值。

2.5.6 ZINCRBY

ZINCRBY 命令的基本语法如下。

```
ZINCRBY key increment member
```

ZINCRBY 用于将有序集合 key 的 member 的 score 值加上增量 increment。也可以通过传递一个负数增量 increment,让 score 值减去相应的值。比如 ZINCRBY key -5 member,就是让 member 的 score 值减去 5。

当 key 不存在,或 member 不是 key 的元素时,ZINCRBY key increment member 等同于 ZADD key increment member。

返回值:member 的新 score 值,以字符串形式表示。

实例如下。

```
127.0.0.1:6379> ZSCORE salary wangwu
"3000"
127.0.0.1:6379> ZINCRBY salary 5000 wangwu
"8000"
127.0.0.1:6379> ZRANGE salary 0 -1 WITHSCORES
1) "lisi"
2) "4000"
3) "zhangsan"
4) "5000"
5) "wangwu"
6) "8000"
```

2.5.7 ZRANGE

ZRANGE 命令的基本语法如下。

```
ZRANGE key start stop [WITHSCORES]
```

ZRANGE 用于返回有序集合 key 中指定区间内的元素。

返回值:在指定区间内,带有 score 值的有序集合 key 的元素的列表。

实例如下。

```
127.0.0.1:6379> ZADD salary2 5000 wangwu
(integer) 1
127.0.0.1:6379> ZADD salary2 10000 lisi
(integer) 1
127.0.0.1:6379> ZADD salary2 3500 zhangsan
(integer) 1

# 显示整个有序集合元素
127.0.0.1:6379> ZRANGE salary2 0 -1 WITHSCORES
```

```
1) "zhangsan"
2) "3500"
3) "wangwu"
4) "5000"
5) "lisi"
6) "10000"
```

我们查看有序集合 salary2,如图 2-3 所示。

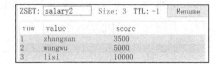

图 2-3 查看有序集合

```
# 显示有序集合索引为 1~2 的元素
127.0.0.1:6379> ZRANGE salary2 1 2 WITHSCORES
1) "wangwu"
2) "5000"
3) "lisi"
4) "10000"

# 测试 stop 超出最大索引时的情况
127.0.0.1:6379> ZRANGE salary2 0 10 WITHSCORES
1) "zhangsan"
2) "3500"
3) "wangwu"
4) "5000"
5) "lisi"
6) "10000"

# 测试当给定区间不存在时的情况
127.0.0.1:6379> ZRANGE salary2 10 20
(empty list or set)
```

2.5.8 ZREVRANGE

ZREVRANGE 命令的基本语法如下。

`ZREVRANGE key start stop [WITHSCORES]`

ZREVRANGE 用于返回有序集合 key 中索引从 start 到 stop 的所有元素。其中元素按 score 值递减（从大到小）排列。

返回值：指定区间内，带有 score 值（可选）的有序集合 key 的元素的列表。

实例如下。

```
# 递增排列有序集合
127.0.0.1:6379> ZRANGE salary 0 -1 WITHSCORES
1) "lisi"
2) "4000"
3) "zhangsan"
4) "5000"
5) "wangwu"
6) "8000"

# 递减排列有序集合
127.0.0.1:6379> ZREVRANGE salary 0 -1 WITHSCORES
1) "wangwu"
2) "8000"
3) "zhangsan"
4) "5000"
```

```
5) "lisi"
6) "4000"
```
从本例可以看出，使用 ZREVRANGE 可以使有序集合 salary 中的元素按 score 值递减排列，再取出全部元素。

2.5.9　ZREVRANGEBYSCORE

ZREVRANGEBYSCORE 命令的基本语法如下。

```
ZREVRANGEBYSCORE key max min [WITHSCORES] [LIMIT offset count]
```

ZREVRANGEBYSCORE 用于返回有序集合 key 中 score 值介于 max 和 min 之间（默认包括 score 值等于 max 或 min）的所有的元素。其中有序集合 key 中的元素按 score 值递减排列。

返回值：指定区间内，带有 score 值的有序集合 key 的元素的列表。

实例如下。

```
127.0.0.1:6379> ZRANGE salary 0 -1 WITHSCORES
1) "lisi"
2) "4000"
3) "zhangsan"
4) "5000"
5) "wangwu"
6) "8000"

# 删除有序集合中 score 值为 4000~5000 的元素
127.0.0.1:6379> ZREMRANGEBYSCORE salary 4000 5000
(integer) 2

127.0.0.1:6379> ZRANGE salary 0 -1 WITHSCORES
1) "wangwu"
2) "8000"
```

在本例中，我们将有序集合 salary 中的元素按 score 值递减排列，并将 score 值为 4000～5000 的元素删除。

2.5.10　ZRANK

ZRANK 命令的基本语法如下。

```
ZRANK key member
```

ZRANK 用于返回有序集合 key 中 member 的排名。其中有序集合 key 中的元素按 score 值递增（从小到大）排列。

返回值：如果 member 是有序集合 key 的元素，则返回 member 的排名；如果 member 不是有序集合 key 的元素，则返回 nil。

实例如下。

```
# 显示有序集合中所有元素及其 score 值
127.0.0.1:6379> ZADD salary3 5000 wangwu
(integer) 1
127.0.0.1:6379> ZADD salary3 10000 lisi
(integer) 1
127.0.0.1:6379> ZADD salary3 4000 zhangsan
(integer) 1
```

```
127.0.0.1:6379> ZRANGE salary3 0 -1 WITHSCORES
1) "zhangsan"
2) "4000"
3) "wangwu"
4) "5000"
5) "lisi"
6) "10000"

# 显示 lisi 的 score 值的排名,按照从小到大的顺序，则其排名第二
127.0.0.1:6379> ZRANK salary3 lisi
(integer) 2
```

2.5.11 ZREVRANK

ZREVRANK 命令的基本语法如下。

`ZREVRANK key member`

ZREVRANK 用于返回有序集合 key 中 member 的排名。其中有序集合 key 中的元素按 score 值递减排列。

排名以 0 为底，也就是说，score 值最大的成员排名为 0。

返回值：如果 member 是有序集合 key 的元素，则返回 member 的排名；如果 member 不是有序集合 key 的元素，则返回 nil。

实例如下。

```
# 添加有序集合元素
127.0.0.1:6379> ZADD salary5 3000 wangwu
(integer) 1
127.0.0.1:6379> ZADD salary5 10000 lisi
(integer) 1
127.0.0.1:6379> ZADD salary5 5000 zhangsan
(integer) 1
127.0.0.1:6379> ZRANGE salary5 0 -1 WITHSCORES
1) "wangwu"
2) "3000"
3) "zhangsan"
4) "5000"
5) "lisi"
6) "10000"

# wangwu 的 score 值排第二
127.0.0.1:6379> ZREVRANK salary5 wangwu
(integer) 2

# lisi 的 score 值最大
127.0.0.1:6379> ZREVRANK salary5 lisi
(integer) 0
```

在本例中，对有序集 salary5 中的元素按 score 值递减排列，lisi 是第一个元素，索引是 0。

2.5.12 ZREMRANGEBYRANK

ZREMRANGEBYRANK 命令的基本语法如下。

ZREMRANGEBYRANK key start stop

ZREMRANGEBYRANK 用于删除有序集合 key 中指定区间内的所有元素。区间以索引参数 start 和 stop 指出，包含 start 和 stop 在内。

索引参数 start 和 stop 都以 0 为底，以 0 表示有序集合 key 的第一个元素，以 1 表示有序集合 key 的第二个元素，其他正数依此类推。也可以使用负数索引，以-1 表示最后一个元素，-2 表示倒数第二个元素，其他负数依此类推。

返回值：被删除元素的数量。

实例如下。

```
127.0.0.1:6379> ZRANGE salary5 0 -1 WITHSCORES
1) "wangwu"
2) "3000"
3) "zhangsan"
4) "5000"
5) "lisi"
6) "10000"

# 删除索引为 0~1 元素
127.0.0.1:6379> ZREMRANGEBYRANK salary5 0 1
(integer) 2

# 显示有序集合内所有元素及其 score 值，有序集合只剩下一个元素
127.0.0.1:6379> ZRANGE salary5 0 -1 WITHSCORES
1) "lisi"
2) "10000"
```

在本例中，我们将有序集合 salary5 中的元素按 score 值递增排列并将索引为 0～1 的元素删除了。

2.5.13 ZREMRANGEBYSCORE

ZREMRANGEBYSCORE 命令的基本语法如下。

ZREMRANGEBYSCORE key min max

ZREMRANGEBYSCORE 用于删除有序集合 key 中所有 score 值介于 min 和 max 之间（默认包括 score 值等于 min 或 max）的元素。

返回值：被删除元素的数量。

实例如下。

```
# 添加有序集合元素
127.0.0.1:6379> ZADD salary6 3000 wangwu
(integer) 1
127.0.0.1:6379> ZADD salary6 10000 lisi
(integer) 1
127.0.0.1:6379> ZADD salary6 5000 zhangsan
(integer) 1

# 显示有序集合内所有元素及其 score 值
127.0.0.1:6379> ZRANGE salary6 0 -1 WITHSCORES
1) "wangwu"
2) "3000"
```

```
3) "zhangsan"
4) "5000"
5) "lisi"
6) "10000"

# 删除所有 score 值为 2500~5500 的元素
127.0.0.1:6379> ZREMRANGEBYSCORE salary6 2500 5500
(integer) 2

# 剩下的有序集合元素
127.0.0.1:6379> ZRANGE salary6 0 -1 WITHSCORES
1) "lisi"
2) "10000"
```

在本例中,我们将有序集合 salary6 中的元素按 score 值递增排列,并将 score 值为 2500~5500 的元素删除了。

2.5.14 ZINTERSTORE

ZINTERSTORE 命令的基本语法如下。

```
ZINTERSTORE destination numkeys key [key ...] [WEIGHTS weight [weight ...]]
[AGGREGATE SUM|MIN|MAX]
```

ZINTERSTORE 用于计算给定的多个有序集合 key 的交集,其中有序集合 key 的数量必须由 numkeys 参数指定,并将该交集(结果集)存储到集合 destination。

返回值:存储到集合 destination 的结果集的基数。

实例如下。

```
127.0.0.1:6379> ZADD test1 70 "Li Lei"
(integer) 1
127.0.0.1:6379> ZADD test1 70 "Han Meimei"
(integer) 1
127.0.0.1:6379> ZADD test1 99.5 "Tom"
(integer) 1

127.0.0.1:6379> ZADD test2 88 "Li Lei"
(integer) 1
127.0.0.1:6379> ZADD test2 75 "Han Meimei"
(integer) 1
127.0.0.1:6379> ZADD test2 99.5 "Tom"
(integer) 1

127.0.0.1:6379> ZINTERSTORE test3 2 test test2
(integer) 3

# 显示有序集合内所有元素及其 score 值
127.0.0.1:6379> ZRANGE test3 0 -1 WITHSCORES
1) "Han Meimei"
2) "145"
3) "Li Lei"
4) "158"
5) "Tom"
6) "199"
```

2.5.15 ZUNIONSTORE

ZUNIONSTORE 命令的基本语法如下。

```
ZUNIONSTORE destination numkeys key [key ...] [WEIGHTS weight [weight ...]]
[AGGREGATE SUM|MIN|MAX]
```

ZUNIONSTORE 用于计算给定的多个有序集合 key 的并集，其中有序集合 key 的数量必须由 numkeys 参数指定，并将该并集（结果集）存储到集合 destination。

WEIGHTS 选项与前面设定的有序集合 key 对应，key 中每一个 score 都要乘以对应的权重。

AGGREGATE 选项指定并集结果的聚合方式：

- SUM：将所有集合中某一个元素的 score 值之和作为结果集中该元素的 score 值。
- MIN：将所有集合中某一个元素的 score 值中最小值作为结果集中该元素的 score 值。
- MAX：将所有集合中某一个元素 score 值中最大值作为结果集中该元素的 score 值。

返回值：存储到集合 destination 的结果集的基数。

实例如下。

```
127.0.0.1:6379> ZADD programmer 2000 peter 3500 jack 5000 tom
(integer) 3
127.0.0.1:6379> ZADD manager 2000 herry 3500 mary 4000 bob
(integer) 3
127.0.0.1:6379> ZRANGE programmer 0 -1 WITHSCORES
1) "peter"
2) "2000"
3) "jack"
4) "3500"
5) "tom"
6) "5000"

127.0.0.1:6379> ZRANGE manager 0 -1 WITHSCORES
1) "herry"
2) "2000"
3) "mary"
4) "3500"
5) "bob"
6) "4000"

# 公司决定给经理（manager）加薪，而不给程序员（programmer）加薪
127.0.0.1:6379> ZUNIONSTORE salary 2 programmer manager WEIGHTS 1 3
(integer) 6

127.0.0.1:6379> ZRANGE salary 0 -1 WITHSCORES
1) "peter"
2) "2000"
3) "jack"
4) "3500"
5) "tom"
6) "5000"
7) "herry"
8) "6000"
```

```
 9) "mary"
10) "10500"
11) "bob"
12) "12000"
```

2.6　Redis HyperLogLog

　　Redis 2.8.9 中添加了 HyperLogLog。Redis 的 HyperLogLog 是用来做基数统计的，主要使用场景是海量数据的计算。HyperLogLog 的优点是，在输入元素的数量非常多时，计算基数所需的空间总是很小。HyperLogLog 只会根据输入元素来计算基数，而不会存储元素本身。基数就是不重复元素的个数。例如数据集{1, 3, 5, 7, 5, 7, 8}，那么这个数据集的基数集为{1, 3, 5 ,7, 8}，基数为 5。HyperLogLog 可以看作一种算法，它提供了不精确的基数计数方案。

　　HyperLogLog 一开始就是为了大数据量的统计而发明的，很适合那种数据量很大，又允许有一点误差的计算，例如页面用户访问量。HyperLogLog 提供了不精确的去重技术方案，标准误差是 0.81%，这对于页面用户访问量的统计是可以接受的。因为访问量可能非常大，但是访问量统计对准确率要求没那么高，没必要做到绝对准确，HyperLogLog 正好符合这种要求，不会占用太多存储空间，同时性能也不错。总之，Redis 的 HyperLogLog 特别适用对海量数据进行统计，对内存占用有要求，并且能够接受一定的错误率的场景。

2.6.1　Redis HyperLogLog 常用命令

HyperLogLog 常用命令及其描述如表 2-1 所示。

表 2-1　　　　　　　　　　HyperLogLog 常用命令及其描述

常用命令	描述
PFADD key element [element ...]	添加指定元素到 HyperLogLog 中
PFCOUNT key [key ...]	返回 HyperLogLog 的基数估算值
PFMERGE destkey sourcekey [sourcekey ...]	将多个 HyperLogLog 合并为一个 HyperLogLog

2.6.2　Redis HyperLogLog 实例

分别统计页面 page1、page2 的用户访客数。
```
# 用户 user1,user2,user3 访问了页面 page1
127.0.0.1:6379> PFADD page1 user1
(integer) 1
127.0.0.1:6379> PFADD page1 user2
(integer) 1
127.0.0.1:6379> PFADD page1 user3
(integer) 1
127.0.0.1:6379> PFCOUNT page1
(integer) 3
# 用户 user3,user4 访问了页面 page2
127.0.0.1:6379> PFADD page2 user3 user4
(integer) 1
127.0.0.1:6379> PFCOUNT page2
```

```
(integer) 2
```
统计两个页面 page1 和 page2 的用户访客数,就需要使用 PFMERGE 命令合并统计了。
```
127.0.0.1:6379> PFMERGE page1-page2 page1 page2
OK
127.0.0.1:6379> PFCOUNT page1-page2
(integer) 4
```
从统计结果可以看出,页面 page1 和 page2 的访客数为 4。

第3章 Redis 常用命令

Redis 提供了丰富的命令，可以对数据库和各种数据类型进行操作，这些命令可以在 Windows 和 Linux 中使用。下面将 Redis 提供的常用命令做一个总结。

3.1 键值相关命令

3.1.1 KEYS

KEYS 用于返回满足 pattern 的所有 key，pattern 支持以下通配符。
- *：匹配任意字符。
- ?：匹配一个任意字符。
- []：匹配方括号内任一单个字符，例如[a-z]表示匹配 26 个小写字母中的任意一个字符，a[b-e]表示匹配 ab、ac、ad 和 ae 字符串。
- \x：匹配特殊字符，例如\?、*。

实例如下。
```
# 添加测试数据
127.0.0.1:6379> MSET one 1 two 2 three 3 four 4 five 5
OK

# 查看当前数据库下的所有 key
127.0.0.1:6379> KEYS *
1) "one"
2) "two"
3) "three"
4) "four"
5) "five"

# 查找第一个字符为 f 的 key
127.0.0.1:6379> KEYS f*
1) "four"
2) "five"

# 查找第一个字符为 t 的 key
127.0.0.1:6379> KEYS t??
```

```
1) "two"

# 查找第二个字符为 o 的 key
127.0.0.1:6379> KEYS ?o*
1) "four"

# 查找以 f 开头的, 包含字母字符串的 key
127.0.0.1:6379> KEYS f[a-z]*
1) "four"
2) "five"
```

使用 KEYS * 可以得到当前 Redis 数据库中的所有 key。

3.1.2 SCAN

SCAN 用于迭代数据库中的 key。SCAN 命令是一个基于游标的迭代器,每次被调用之后都会向用户返回一个新游标,用户在下次迭代时需要使用这个新游标作为 SCAN 命令的游标参数,以此来延续之前的迭代过程。

SCAN 返回一个包含两个元素的数组,第一个元素是用于进行下一次迭代的新游标,而第二个元素则是一个数组,这个数组中包含了所有被迭代的元素。如果返回的新游标为 0 则表示迭代已结束。

SCAN 命令的基本语法如下。

```
SCAN cursor [MATCH pattern] [COUNT count]
```

SCAN 命令的参数包括:

- cursor: 游标。
- pattern: 匹配的模式。
- count: 指定从数据集里返回多少元素, 默认值为 10。

实例如下。

使用 SET 命令创立 50 条 String 类型的数据。限于篇幅本节只列出创建 2 条数据的命令,剩下 48 条数据,请读者自行创建。

```
127.0.0.1:6379> SET key:1 1
OK
127.0.0.1:6379> SET key:2 2
OK
```

使用 SCAN 命令迭代数据库中的 key。

```
127.0.0.1:6379> SCAN 0           # 使用 0 作为游标, 开始新的迭代
1) "20"                          # 第 1 次迭代时返回的游标
2)  1) "key:40"
    2) "key:49"
    3) "key:48"
    4) "key:24"
    5) "key:34"
    6) "key:3"
    7) "key:37"
    8) "key:5"
    9) "key:32"
   10) "key:33"
```

```
    11) "key:7"
127.0.0.1:6379> SCAN 20    # 使用第一次迭代时返回的游标 17 开始新的迭代
1) "26"                    # 第 2 次迭代时返回的游标
2)  1) "key:42"
    2) "key:13"
    3) "key:30"
    4) "key:20"
    5) "key:38"
    6) "key:25"
    7) "key:43"
    8) "key:14"
    9) "key:47"
    10) "key:28"
```

3.1.3　EXISTS

EXISTS 用于查看 key 是否存在，如果 key 存在则返回 1，否则返回 0。

实例如下。

```
127.0.0.1:6379> SET name xinping
OK
127.0.0.1:6379> EXISTS name
(integer) 1
127.0.0.1:6379> EXISTS address
(integer) 0
```

结果表明不存在 address，但是存在 name。

3.1.4　DEL

DEL 用于删除 key，返回被删除 key 的个数。

实例如下。

```
127.0.0.1:6379> DEL name
(integer) 1
127.0.0.1:6379> DEL name
(integer) 0
```

在本例中可以看出 name 是存在的，删除后就不存在了。

3.1.5　EXPIRE

EXPIRE 用于设置 key 的过期时间，单位为秒。超过该时间后，key 被自动删除。

返回值为 1 表示已设置 key 的过期时间；返回值为 0 则表示 key 不存在，不能设置其过期时间。

　　如果 key 已经存在过期时间，则通过 EXPIRE 设置的时候会覆盖之前的过期时间。

实例如下。

```
127.0.0.1:6379> SET cache_page "www.jd.com"
OK
127.0.0.1:6379> EXPIRE cache_page 60
```

```
(integer) 1
```
在本例中,使用 EXPIRE 让 cache_page 存在 60s。等待 60s 后 cache_page 被自动删除,就不存在了。可以使用 EXISTS 查看 cache_page。
```
127.0.0.1:6379> EXISTS cache_page
(integer) 0
```

3.1.6 TTL

TTL 用于获取 key 所剩的过期时间。该命令以秒为单位返回 key 的剩余时间,如果 key 不存在或没有超时设置,则返回-2。

实例如下。
```
127.0.0.1:6379> SET cache_page2 "www.jd.com"
OK
127.0.0.1:6379> EXPIRE cache_page2 60
(integer) 1
127.0.0.1:6379> TTL cache_page2
(integer) 56
127.0.0.1:6379> TTL cache_page2
(integer) 35
127.0.0.1:6379> TTL cache_page2
(integer) 25
127.0.0.1:6379> TTL cache_page2
(integer) 8
127.0.0.1:6379> TTL cache_page2
(integer) -2
```
在本例中,我们设置 cache_page2 的过期时间是 60s,然后我们不断用 TTL 来获取 cache_page2 的剩余时间,直至为-2,说明 cache_page2 已过期。

3.1.7 SELECT

SELECT 用于选择数据库,数据库为 0~15(一共 16 个数据库)。

实例如下。
```
SELECT 1
```
在本例中,选择数据库 1。

3.1.8 MOVE

MOVE,用于将当前数据库中的 key 转移到其他数据库中。

实例如下。
```
127.0.0.1:6379[1]> SELECT 0
OK
127.0.0.1:6379> SET age 20
OK
127.0.0.1:6379> GET age
"20"
127.0.0.1:6379> MOVE age 1
(integer) 1
127.0.0.1:6379> GET age
(nil)
```

```
127.0.0.1:6379[1]> SELECT 1
OK
127.0.0.1:6379[1]> GET age
"20"
```
在本例中,我们先显式地选择了数据库 0,然后在这个数据库中设置一个 key,即 age;接下来我们将 age 从数据库 0 转移到数据库 1;然后我们在数据库 0 中确认了没有 age,但在数据库 1 中存在 age,说明转移 age 成功了。

3.1.9　PERSIST

PERSIST 用于删除 key 的过期时间。

实例如下。
```
127.0.0.1:6379> SET age 20
OK
127.0.0.1:6379> EXPIRE age 20
(integer) 1
127.0.0.1:6379> TTL age
(integer) 18
127.0.0.1:6379> PERSIST age
(integer) 1
127.0.0.1:6379> PERSIST age
(integer) 0
127.0.0.1:6379> TTL age
(integer) -1
```
在本例中,我们手动地删除了 age 的过期时间。

3.1.10　RANDOMKEY

RANDOMKEY 用于随机返回 key 空间中的一个 key。

实例如下。
```
127.0.0.1:6379> SET age1 20
OK
127.0.0.1:6379> SET age2 21
OK
127.0.0.1:6379> RANDOMKEY
"age2"
127.0.0.1:6379> RANDOMKEY
"age1"
```
通过本例的结果可以看到,取 key 时是随机的。

3.1.11　RENAME

RENAME 用于重命名 key。

实例如下。
```
127.0.0.1:6379> SET age3 20
OK
127.0.0.1:6379> RENAME age3 age4
OK
127.0.0.1:6379> GET age4
```

```
"20"
```
在本例中，我们看到 age3 被我们成功改名为 age4 了。

3.1.12　TYPE

TYPE 用于获取 key 关联值的类型，并以字符串的格式返回结果。返回的字符串为 String、List、Set、Hash 和 Sorted Set，如果 key 不存在则返回 none。

实例如下。
```
127.0.0.1:6379> SET name wangwu
OK
127.0.0.1:6379> TYPE name
string
127.0.0.1:6379> LPUSH ls1 a
(integer) 1
127.0.0.1:6379> TYPE ls1
List
127.0.0.1:6379> TYPE age
none
```
在本例中，可以看出使用 TYPE 命令能返回 key 关联值的类型。

3.2　服务器相关命令

3.2.1　PING

PING 使用客户端向 Redis 服务器发送一个 "PING" 字符串。如果 Redis 服务器运行正常的话，会返回一个 "PONG" 字符串，用来测试客户端与 Redis 服务器的连接是否依然生效。

实例如下。
```
127.0.0.1:6379> PING
PONG
```

3.2.2　ECHO

可以通过 ECHO 在命令行输出一些内容。

实例如下。
```
127.0.0.1:6379> ECHO "hello world"
"hello world"
```

3.2.3　QUIT

可以通过 QUIT 退出当前 Redis 连接。

实例如下。
```
127.0.0.1:6379> QUIT
[root@bogon bin]#
```

3.2.4　DBSIZE

DBSIZE 用于查看当前数据库中 key 的数目。

实例如下。
```
127.0.0.1:6379> DBSIZE
(integer) 2
127.0.0.1:6379> KEYS *
1) "name"
2) "ls1"
```
在本例中，可以看出当前数据库中有两个 key。

3.2.5 INFO

INFO 用于查看 Redis 服务器的各种信息和统计数值。

实例如下。
```
127.0.0.1:6379> INFO
# Server                           #Redis 的服务器信息
redis_version:6.0.6
redis_git_sha1:00000000
redis_git_dirty:0
redis_build_id:b50d302201129968
redis_mode:standalone
os:Linux 3.10.0-693.11.6.el7.x86_64 x86_64
arch_bits:64
multiplexing_api:epoll             # Redis 的事件循环机制
atomicvar_api:atomic-builtin
gcc_version:4.8.5
process_id:2800
run_id:4dd2e989b43c18f63d32eab9745d98abe00929ae  # 标识 Redis 服务器的随机值
tcp_port:6379
uptime_in_seconds:7                # Redis 服务器启动的时间(单位 s)
uptime_in_days:0                   # Redis 服务器启动的时间(单位 day)
hz:10
lru_clock:6416788
executable:/root/redis-server
config_file:/usr/local/redis/conf/redis.conf

# Clients                          # 已连接客户端信息
connected_clients:1                # 连接的客户端数
client_longest_output_list:0       # 当前客户端连接的最大输出列表
client_biggest_input_buf:0         # 当前客户端连接的最大输入 buffer
blocked_clients:0                  # 被阻塞的客户端数

# Memory                           # 内存信息
used_memory:828512                 # 使用内存(单位 Byte)
used_memory_human:809.09K          # 以更直观的单位显示分配的内存总量
used_memory_rss:4763648            # 系统给 Redis 分配的内存(即常驻内存)
used_memory_rss_human:4.54M
used_memory_peak:828512            # 内存使用的峰值大小
used_memory_peak_human:809.09K     # 以更直观的单位显示内存使用峰值
used_memory_peak_perc:100.13%
used_memory_overhead:815278
used_memory_startup:765648
```

```
used_memory_dataset:13234
used_memory_dataset_perc:21.05%
total_system_memory:1022627840
total_system_memory_human:975.25M
used_memory_lua:37888                    # Lua 引擎使用的内存
used_memory_lua_human:37.00K
maxmemory:0
maxmemory_human:0B
maxmemory_policy:noeviction
mem_fragmentation_ratio:5.75             # used_memory_rss/used_memory 比例
mem_allocator:jemalloc-4.0.3             # 内存分配器
active_defrag_running:0
lazyfree_pending_objects:0

# Persistence                            # 持久化的相关信息
loading:0
rdb_changes_since_last_save:0            # 自上次 RDB 保存以后更改的次数
rdb_bgsave_in_progress:0                 # 表示当前是否在进行 bgsave 操作，如果是则为 1
rdb_last_save_time:1516366221            # 上次保存 RDB 文件的时间戳
rdb_last_bgsave_status:ok                # 上次保存的状态
rdb_last_bgsave_time_sec:-1              # 上次保存 RDB 文件已花费的时间 (单位 s)
rdb_current_bgsave_time_sec:-1           # 目前保存 RDB 文件已花费的时间（单位 s）
rdb_last_cow_size:0
aof_enabled:0                            # 是否开启 AOF，默认没开启
aof_rewrite_in_progress:0                # 标识 AOF 的 rewrite 操作是否在进行
aof_rewrite_scheduled:0                  # 标识是否将要在 RDB 保存操作结束后执行
aof_last_rewrite_time_sec:-1             # 上次 rewrite 操作使用的时间 (单位 s)
aof_current_rewrite_time_sec:-1          # 如果 rewrite 操作正在进行，则记录所使用的时间
aof_last_bgrewrite_status:ok             # 上次 rewrite 操作的状态
aof_last_write_status:ok
aof_last_cow_size:0
# 开启 AOF 后增加的一些信息
aof_current_size:0                       # AOF 当前大小
aof_base_size:0                          # AOF 上次启动或 rewrite 的大小
aof_pending_rewrite:0                    # 同上面的 aof_rewrite_scheduled
aof_buffer_length:0                      # aof buffer 的大小
aof_rewrite_buffer_length:0              # aof rewrite buffer 的大小
aof_pending_bio_fsync:0                  # 后台 I/O 队列中等待 fsync 任务的个数
aof_delayed_fsync:0                      # 延迟的 fsync 计数器 TODO
------------------------------
# Stats                                  # 一般统计信息
total_connections_received:1             # 自启动起连接过的总数
total_commands_processed:1               # 自启动起执行命令的总数
instantaneous_ops_per_sec:0              # 每秒执行的命令个数
total_net_input_bytes:31
total_net_output_bytes:10163
instantaneous_input_kbps:0.01
instantaneous_output_kbps:6.14
rejected_connections:0                   # 因为最大客户端连接限制而导致被拒绝连接的个数
sync_full:0
sync_partial_ok:0
```

```
sync_partial_err:0
expired_keys:0                          # 自启动起过期的 key 的个数
evicted_keys:0                          # 因为内存大小限制而被驱逐出去的 key 的个数
keyspace_hits:0                         # 在 main dictionary(todo) 中成功查到的 key 的个数
keyspace_misses:0                       # 在 main dictionary(todo) 中未查到的 key 的个数
pubsub_channels:0                       # 发布/订阅频道数
pubsub_patterns:0                       # 发布/订阅模式数
latest_fork_usec:0                      # 上次的 fork 操作使用的时间 (单位 ms)
migrate_cached_sockets:0
slave_expires_tracked_keys:0
active_defrag_hits:0
active_defrag_misses:0
active_defrag_key_hits:0
active_defrag_key_misses:0

# Replication                           # 主/从复制信息
role:master                             # 角色
connected_slaves:0                      # 连接的从库数
master_replid:4066f05f8e6cfcf9228a1450a25efaa551f954a2
master_replid2:0000000000000000000000000000000000000000
master_repl_offset:0
second_repl_offset:-1
repl_backlog_active:0
repl_backlog_size:1048576
repl_backlog_first_byte_offset:0
repl_backlog_histlen:0

# CPU                                   # CPU 计算量的统计信息
used_cpu_sys:0.01                       # Redis 服务器的系统 CPU 使用率
used_cpu_user:0.00                      # Redis 服务器的用户 CPU 使用率
used_cpu_sys_children:0.00              # 后台进程的系统 CPU 使用率
used_cpu_user_children:0.00             # 后台进程的用户 CPU 使用率

# Cluster                               # Redis 集群信息
cluster_enabled:0

# Keyspace                              # 数据库相关统计信息
db0:keys=1,expires=0,avg_ttl=0          # 数据库 0 中的 key 的个数
```

3.2.6 MONITOR

MONITOR 用于实时输出 Redis 服务器接收到的命令，可供调试使用。

首先使用 redis-cli 命令打开第一个客户端，执行以下命令。

```
127.0.0.1:6379> MONITOR
OK
```

然后使用 redis-cli 命令打开第二个客户端，执行以下命令存储数据。

```
127.0.0.1:6379> SET name wangwu
OK
127.0.0.1:6379> GET name
"wangwu"
```

最后查看第一个客户端，会看到如下信息。

```
127.0.0.1:6379> MONITOR
OK
1525610398.977423 [0 127.0.0.1:53469] "set" "name" "wangwu"
1525610400.361838 [0 127.0.0.1:53469] "get" "name"
```

从结果可看出，此 Redis 服务器目前接收了命令 set 和 get。

3.2.7　CONFIG GET

CONFIG GET 用于获取 Redis 服务器的配置信息。

实例如下。

```
127.0.0.1:6379> CONFIG GET dir
1) "dir"
2) "/usr/local/redis/bin"
```

在本例中，我们使用 CONFIG GET 获取了 dir 参数配置的值。如果想获取全部参数配置的值，执行 CONFIG GET *就可以将全部参数配置的值都显示出来。

3.2.8　FLUSHDB

FLUSHDB 用于删除当前选择的数据库中的所有 key。

实例如下。

```
127.0.0.1:6379> SELECT 0
OK
127.0.0.1:6379> KEYS *
1) "name"
2) "age"
127.0.0.1:6379> FLUSHDB
OK
```

在本例中，我们将数据库 0 中的 key 都删除了。

3.2.9　FLUSHALL

FLUSHALL 用于删除所有数据库中的 key。

实例如下。

```
127.0.0.1:6379> FLUSHALL
OK
127.0.0.1:6379> KEYS *        #查询数据库 0 的所有 key
(empty list or set)
127.0.0.1:6379> SELECT 1      #使用数据库 1
OK
127.0.0.1:6379[1]> KEYS *     #查询数据库 1 的所有 key
(empty list or set)
```

在本例中，我们使用 FLUSHALL 命令删除了所有数据库中的 key，查看数据库 0 中的 key，发现都被删除了；然后切换到数据库 1，发现数据库 1 中的 key 也被删除了。

第 4 章　Redis 高级主题

本章主要介绍 Redis 的高级主题，包括服务器配置、Redis 事务、Redis 发布和订阅、Pipeline 批量发送请求、数据备份与恢复等。

4.1　服务器配置

在 Windows 和 Linux 的 Redis 服务器里面，都有一个配置文件。Redis 配置文件位于 Redis 安装目录下，在不同操作系统下，Redis 配置文件名是不一样的。

- 在 Windows 下，Redis 配置文件名为 redis.windows.conf。
- 在 Linux 下，Redis 配置文件名为 redis.conf。

在 Linux 下使用 redis-server 命令启动 Redis 服务器时，可以在命令后面指定配置文件。例如，在 Linux 下使用如下命令启动 Redis 服务器。

```
$ redis-server /usr/local/redis/conf/redis.conf
```

也可以通过 Redis 的 CONFIG 命令对 Redis 配置文件进行查看或设置项。

4.1.1　Redis 服务器允许远程主机访问

若主机需要远程访问 Redis 服务器，则可以修改 Redis 配置文件 redis.conf。文件中 bind 字段默认为 bind 127.0.0.1，表示只能在本机访问 Redis 服务器，如图 4-1 所示。

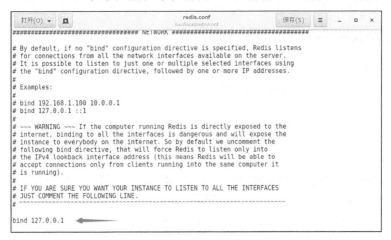

图 4-1　需修改 redis.conf 配置文件的 bind 字段以允许远程主机访问 Redis 服务器

若允许远程主机访问 Redis 服务器,可以将 bind 127.0.0.1 改为 bind 0.0.0.0。

找到 daemonize no,把这一项的 no 改成 yes。daemonize yes 主要是开启 Redis 的守护进程,可以在 Linux 启动时自动运行 Redis,将 Redis 服务器作为守护进程(daemon)来运行。

```
daemonize yes
```

关闭 protected-mode 模式,让外部网络可以直接访问 Redis 服务器。在 Redis 配置文件中找到 protected-mode yes,把这一项的 yes 改成 no。

```
protected-mode no
```

4.1.2 客户端远程连接 Redis 服务器

远程连接 Redis 服务器需要使用 redis-cli 命令,redis-cli 命令的用法为:redis-cli [OPTIONS] [cmd [arg [arg ...]]]。

redis-cli 命令的关键参数如下。

- -h:主机 IP 地址,默认是 127.0.0.1。
- -p:端口,默认是 6379。
- -a:密码,如果 Redis 设置了密码,需要传递密码。

假设有两台 Redis 服务器,Redis 服务器的 IP 地址分别是 192.168.1.11 和 192.168.1.14。现在需要在 192.168.1.11 上通过 redis-cli 命令远程访问 192.168.1.14 上的 Redis 服务器,在 192.168.1.11 上通过以下命令远程连接 192.168.1.14 上的 Redis 服务器。

```
$ redis-cli -h 192.168.1.14 -p 6379
192.168.1.14:6379> SET age 20
OK
192.168.1.14:6379> GET age
"20"
```

通过本例,客户端远程连接到 192.168.1.14 上的 Redis 服务器,设置了一个 String 类型的值,使 age 等于 "20"。

4.1.3 设置密码

通过 Redis 配置文件来设置密码,客户端连接到 Redis 服务器时就需要密码验证,这样可以让 Redis 服务器更安全。

1. 通过命令设置 Redis 密码

首先通过 CONFIG GET requirepass 命令查看 Redis 服务器是否设置了密码验证。

```
127.0.0.1:6379> CONFIG GET requirepass
1) "requirepass"
2) ""
```

默认情况下 requirepass 对应的参数是空的,即没有密码,表示不需要通过密码验证就可以连接到 Redis 服务器。

然后设置 Redis 服务器的当前密码为 123。Redis 服务器重新启动后又会将密码重置为默认,即没有密码,因此不建议用此种方式设置密码。

```
127.0.0.1:6379> CONFIG SET requirepass "123"
OK
127.0.0.1:6379> CONFIG GET requirepass
```

```
1) "requirepass"
2) "123"
```
通过命令设置 Redis 密码的完整实例如下。
```
[root@bogon bin]# ./redis-cli -h 127.0.0.1 -p 6379
# 获取现有 Redis 服务器密码
127.0.0.1:6379> CONFIG GET requirepass
1) "requirepass"
2) ""
# 设置新密码
127.0.0.1:6379> CONFIG SET requirepass "123"
OK
127.0.0.1:6379> exit
# 重新设置密码后，连接 Redis 服务器需添加密码参数
[root@bogon bin]# ./redis-cli -h 127.0.0.1 -p 6379 -a 123
127.0.0.1:6379> set age 20
OK
```
重新设置密码后，需要重新登录 Redis 服务器才能获取操作权限。

2．通过修改 Redis 配置文件设置密码

编辑 redis.conf 配置文件，添加 requirepass 值设置密码。

requirepass 123

以上设置的 Redis 服务器的访问密码为 123，读者可以根据实际需要进行密码的设置。

重启 Redis 服务器后，再次进入 Redis 客户端并输入以下内容。
```
127.0.0.1:6379>KEYS *
(error) NOAUTH Authentication required
```
发现没有权限进入当前数据库，需要使用 AUTH 命令进行授权操作。
```
127.0.0.1:6379>AUTH 123
OK
```
输入密码则成功进入当前数据库，用这种方式每次进入当前数据库的时候都需要输入密码。

还有一种简单的方式，即直接登录数据库并授权。
```
[root@bogon bin]# ./redis-cli -h 127.0.0.1 -p 6379 -a 123
```
在本例中，我们设置了 Redis 服务器的访问密码为 123。建议用此种方式设置密码，重启 Redis 服务器后就可以使用设置好的密码连接 Redis 服务器了。

4.1.4　Redis 端口修改

Redis 默认的端口是 6379，在 redis.conf 配置文件里搜索 6379 就能找到端口参数配置，如下所示。
```
# Accept connections on the specified port, default is 6379 (IANA #815344).
# If port 0 is specified Redis will not listen on a TCP socket.
port 6379
```
可以将默认的端口修改成指定的端口，不要端口冲突就行。启动 Redis 服务器后，可以使用 ps -ef | grep redis 命令查看 Redis 服务器占用的端口，如图 4-2 所示。

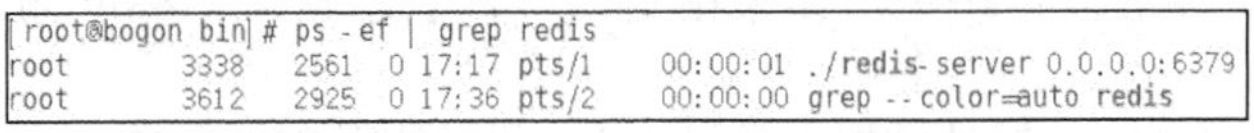

```
[root@bogon bin]# ps -ef | grep redis
root      3338  2561  0 17:17 pts/1    00:00:01 ./redis-server 0.0.0.0:6379
root      3612  2925  0 17:36 pts/2    00:00:00 grep --color=auto redis
```

图 4-2　查看 Redis 服务器占用的端口

从图 4-2 可以看出，现在 Redis 服务器使用的端口是 6379。

4.1.5 查看配置

可以通过 CONFIG GET 命令查看配置。

CONFIG GET 命令的基本语法如下。

```
redis 127.0.0.1:6379> CONFIG GET CONFIG_SETTING_NAME
```

实例：查看 Redis 的日志级别。

```
127.0.0.1:6379> CONFIG GET loglevel
1) "loglevel"
2) "notice"
```

4.1.6 修改配置

可以通过修改 redis.conf 配置文件或使用 CONFIG SET 命令来修改配置。

CONFIG SET 命令的基本语法如下。

```
redis 127.0.0.1:6379> CONFIG SET CONFIG_SETTING_NAME NEW_CONFIG_VALUE
```

实例：修改 Redis 记录的日志级别。

```
127.0.0.1:6379> CONFIG SET loglevel "notice"
OK
127.0.0.1:6379>  CONFIG GET loglevel
1) "loglevel"
2) "notice"
127.0.0.1:6379>
```

4.1.7 配置项说明

redis.conf 配置项说明如下。

（1）Redis 默认不是以守护进程的方式运行的，可以通过 daemonize 配置项修改。如果指定为 yes，表示启用守护进程。

```
daemonize no
```

（2）当 Redis 以守护进程的方式运行时，Redis 默认会把 pid 写入/var/run/redis.pid 文件，可以通过 pidfile 配置项来指定。

```
pidfile /var/run/redis.pid
```

（3）指定 Redis 监听端口，可以通过 port 配置项来指定，默认端口为 6379。

```
port 6379
```

（4）指定绑定的主机地址，可以用于限制连接，默认只能本机访问 Redis 服务器。

```
bind 127.0.0.1
```

（5）指定 Redis 客户端闲置多长时间后关闭连接，如果指定为 0，表示关闭该功能。

```
timeout 300
```

（6）指定日志的记录级别，Redis 支持 4 个级别，即 debug、verbose、notice、warning，默认值为 verbose。

```
loglevel verbose
```

（7）指定日志文件的保存路径，如果指定为" "，表示默认为标准输出。

```
logfile ""
```

（8）指定数据库的数量，默认数据库数量为 0，可以使用 SELECT 命令来连接指定的数据库。

```
databases 16
```
（9）指定在多长时间内执行多少次更新操作，Redis 就将数据同步到数据文件中。
```
save <seconds> <changes>
```
Redis 的默认配置文件中设置了 3 个触发条件。
```
save 900 1

save 300 10

save 60 10000
```
save 900 1 表示 900s（15min）内有 1 个更新，save 300 10 表示 300s（5min）内有 10 个更新，save 60 10000 表示 60s（1min）内有 10 000 个更新，Redis 就将数据同步到数据文件中。

（10）对于存储到磁盘中的 Redis 快照，可以设置是否进行压缩存储，默认值为 yes。如果指定为 yes，Redis 会采用 LZF 压缩算法对存储到磁盘中的 Redis 快照进行压缩。如果不想消耗 CPU 来压缩快照的话，可以指定为 no 来关闭该选项，但是存储在磁盘上的快照会比较大。
```
rdbcompression yes
```
（11）指定存储数据的本地数据库的文件名，默认值为 dump.rdb。
```
dbfilename dump.rdb
```
（12）指定本地数据库存放目录。
```
dir/usr/local/redis/bin
```
（13）指定当本机为 Slave（从服务器）服务时，Master（主服务器）服务的 IP 地址及端口。在 Redis 启动时，它会自动从 Master 服务进行数据同步。
```
slaveof <masterip> <masterport>
```
（14）指定当 Master 服务设置了密码保护时，Slave 服务连接 Master 服务的密码。
```
masterauth <master-password>
```
（15）指定 Redis 的连接密码。如果配置了连接密码，则客户端在连接 Redis 时需要使用 AUTH 命令提供密码，默认配置是关闭的。
```
requirepass foobared
```
（16）指定同一时间内客户端允许的最大连接数，如果设置为 maxclients 0，表示不做限制。当客户端连接数达到最大限制时，Redis 会关闭新的连接并向客户端发送 max number of clients reached 错误信息。
```
maxclients 128
```
（17）指定 Redis 的最大内存限制。Redis 在启动时会把数据缓存到内存中，达到最大内存后，Redis 会尝试清除已到期的 key。Redis 新的 vm（虚拟内存）机制，会把 key 存放在内存，把 value 存放在 swap 区。
```
maxmemory <bytes>
```
（18）指定 Redis 是否在每次执行更新操作后进行日志记录，在默认情况下 Redis 是异步地把内存中的数据写入磁盘的。如果不开启此选项，可能会在主机断电时丢失一段时间内的数据。该选项默认值为 no。
```
appendonly no
```
（19）指定更新日志文件名，默认值为 appendonly.aof。
```
appendfilename appendonly.aof
```
（20）指定更新日志的条件，共有 3 个可选值。
- no：表示等操作系统进行数据缓存后才同步到磁盘，特点是速度快。

- always：表示每次执行更新操作后，需要手动调用 fsync()将数据写到磁盘，特点是速度慢，比较安全。
- everysec：表示每秒同步一次数据到磁盘，是上面两个选项的折中选项，也是默认值。

```
appendfsync everysec
```

（21）指定是否启用虚拟内存机制，默认值为 no。

```
vm-enabled no
```

（22）指定虚拟内存文件路径，默认值为/tmp/redis.swap。不可多个 Redis 实例共享。

```
vm-swap-file /tmp/redis.swap
```

（23）指定将所有大于 vm-max-memory 的数据存入虚拟内存，默认值为 0，无论 vm-max-memory 多小，所有索引数据都是内存存储的（Redis 的索引数据就是 key），也就是说，当 vm-max-memory 设置为 0 的时候，其实是所有 value 都存储在磁盘。

```
vm-max-memory 0
```

（24）Redis 的 swap 文件被分成了很多个 page，一个对象可以保存在多个 page 上，但一个 page 不能被多个对象共享。

```
vm-page-size 32
```

（25）指定 swap 文件中的 page 数量。

```
vm-pages 134217728
```

（26）指定访问 swap 文件的线程数，此选项值最好不要超过计算机的核数。如果设置为 0，那么所有对 swap 文件的操作都是串行的，可能会造成比较长时间的延迟。此选项默认值为 4。

```
vm-max-threads 4
```

（27）指定在向客户端响应时，是否把较小的包合并为一个包发送，默认值为 yes。

```
glueoutputbuf yes
```

4.2 Redis 事务

Redis 的事务可以一次执行多个命令，有以下两个重要的特点。
- 事务是一个单独的隔离操作：事务中的所有命令都会按顺序执行，事务在执行的过程中，不会被其他客户端发送的命令所打断。
- 事务是一个原子性操作：事务中的命令要么全部被执行，要么全部都不执行。

一个事务会经历 3 个阶段，即开始事务、命令入队、执行事务，如图 4-3 所示。

图 4-3　事务的 3 个阶段

4.2.1　Redis 事务的常用命令

表 4-1 列出了 Redis 事务的常用命令及其描述。

表 4-1　Redis 事务的常用命令及其描述

常用命令	描述
DISCARD	取消事务，取消执行事务块内的所有命令
EXEC	执行所有事务块内的命令

常用命令	描述
MULTI	标记一个事务块的开始
UNWATCH	取消 WATCH 命令对所有 key 的监视
WATCH	监视所有的 key，如果在事务执行之前这些 key 被其他命令所改动，那么事务将被打断

4.2.2 简单事务控制

以下是一个简单事务控制的实例。先使用 MULTI 命令开始一个事务，然后将多个命令入队到事务中，最后由 EXEC 命令触发事务，一起执行事务块内的所有命令。

```
127.0.0.1:6379> SET age 20
OK
127.0.0.1:6379> MULTI
OK
#两个SET命令都没有被执行，而是被放到了队列中
127.0.0.1:6379> SET age 21
QUEUED
127.0.0.1:6379> SET age 22
QUEUED
127.0.0.1:6379> GET age
QUEUED
#SADD命令也没有被执行，也被放到了队列中
127.0.0.1:6379> SADD tag "java" "python" "c"
QUEUED
127.0.0.1:6379> SMEMBERS tag
QUEUED
# 触发事务，一起执行队列中的命令，执行命令后返回的是执行 3 个命令的结果
127.0.0.1:6379> EXEC
1) OK
2) OK
3) "22"
4) (integer) 3
5) 1) "python"
   2) "c"
   3) "java"
# 查看 age 对应的字符串
127.0.0.1:6379> GET age
"22"
# 查看 tag 对应的集合
127.0.0.1:6379> SMEMBERS tag
1) "python"
2) "c"
3) "java"
```

从本例中可以看到，两个 SET 命令和一个 SADD 命令并没有被立即执行而是放到了队列中，在执行 EXEC 命令后 3 个命令才被连续执行，最后返回的是 3 个命令的执行结果。

4.2.3 取消一个事务

我们可以执行 DISCARD 命令来取消一个事务，让事务回滚。实例如下。

```
127.0.0.1:6379> SET age 20
OK
127.0.0.1:6379> GET age
"20"
127.0.0.1:6379> MULTI
OK
127.0.0.1:6379> SET age 21
QUEUED
127.0.0.1:6379> SET age 22
QUEUED
127.0.0.1:6379> DISCARD
OK
127.0.0.1:6379> GET age
"20"
```

从本例中可以看到，两个 SET 命令并没有被立即执行而是放到了队列中，在执行 DISCARD 命令后清空事务的命令队列并退出事务的上下文，也就是事务回滚。

4.2.4 乐观锁控制复杂事务

乐观锁就是利用版本号比较机制，只是在读数据的时候，将读到的数据的版本号一起读出来，当对数据的读操作结束并准备写数据的时候，再进行一次数据的版本号的比较。若版本号没有变化，即认为数据是一致的，没有更改，可以直接写入；若版本号有变化，则认为数据被更新，不能写入，防止脏写。

乐观锁工作机制：执行 WATCH 命令监视 Redis 给定的每一个 key，当执行 EXEC 命令时如果监视的任何一个 key 自从执行 WATCH 命令后发生过变化，则整个事务会回滚，不执行任何操作。

Redis 的乐观锁实例测试如表 4-2 所示。

表 4-2　　　　　　　　　　　Redis 的乐观锁实例测试

客户端 1	客户端 2	说明
127.0.0.1:6379> SET age 21 OK 127.0.0.1:6379> SET name zhangsan OK 127.0.0.1:6379> GET age "21" 127.0.0.1:6379> GET name "zhangsan"	127.0.0.1:6379> GET age "21" 127.0.0.1:6379> GET name "zhangsan"	在 Redis 数据库中使用两个客户端登录，并设置 key 的初始值
127.0.0.1:6379> MULTI OK 127.0.0.1:6379> INCR age QUEUED 127.0.0.1:6379> SET name lisi QUEUED		客户端 1 开启事务，并提交命令： 1. 将当前 age 自增 1； 2. 将 name 值改为 lisi
	127.0.0.1:6379> INCR age (integer) 22	客户端 2 修改了 age 值

续表

客户端 1	客户端 2	说明
127.0.0.1:6379> EXEC 1) (integer) 23 2) OK 3) "23" 4) "23" 127.0.0.1:6379> GET age "23" 127.0.0.1:6379> GET name "lisi"		客户端 1 执行事务队列命令，发现 age 不是 22，而是 23，原因是被其他客户端抢先修改了；name 值也修改了。这样可能导致数据不一致

为了解决这个问题引入了乐观锁，使用乐观锁的实例如表 4-3 所示。

表 4-3　　　　　　　　　　　　使用乐观锁的实例

客户端 1	客户端 2	说明
127.0.0.1:6379> FLUSHDB OK 127.0.0.1:6379> SET name zhangsan OK 127.0.0.1:6379> SET age 21 OK 127.0.0.1:6379> GET name "zhangsan" 127.0.0.1:6379> GET age "21"	127.0.0.1:6379> GET name "zhangsan" 127.0.0.1:6379> GET age "21"	在 Redis 数据库中使用两个客户端登录，并设置 key 的初始值
127.0.0.1:6379> WATCH age name OK 127.0.0.1:6379> MULTI OK 127.0.0.1:6379> INCR age QUEUED 127.0.0.1:6379> SET name lisi QUEUED		客户端 1 用 WATCH 命令监视 age 和 name，然后开启事务，并提交队列命令
	127.0.0.1:6379> INCR age (integer) 22	客户端 2 修改了 age 值
127.0.0.1:6379> EXEC (nil) 127.0.0.1:6379> GET name "zhangsan" 127.0.0.1:6379> GET age "22"		客户端 1 执行事务队列命令，使用 WATCH 命令监控发现此期间 age 的值已经被修改过，则让整个事务回滚，不执行任何操作。 WATCH 可以同时监控多个 key，在监控期间只要有一个 key 被其他客户端修改，则整个事务回滚

通过以上基于 Redis 的乐观锁的简单实例，我们可以得出以下两点结论。
（1）乐观锁的实现，必须基于 WATCH 命令，然后利用 Redis 的事务。

（2）WATCH 的生命周期只和事务关联，一个事务执行完毕，相应的 WATCH 的生命周期就会结束。

4.3 Redis 发布和订阅

Redis 发布和订阅是一种消息通信模式：发送者（Publish）用来发送消息，订阅者（Subscribe）用来接收消息。Redis 客户端可以订阅任意数量的频道。

图 4-4 展示了频道 channel1，以及订阅这个频道的 3 个客户端 client2、client5 和 client1 之间的关系。

当有新消息通过 PUBLISH 命令发送给频道 channel1 时，这个消息就会被发送给订阅它的 3 个客户端（client2、client5 和 client1），如图 4-5 所示。

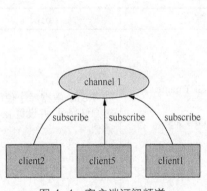

图 4-4　客户端订阅频道

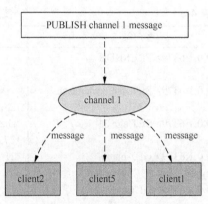

图 4-5　客户端收到消息

4.3.1 Redis 发布和订阅的常用命令

表 4-4 列出了 Redis 发布和订阅的常用命令及其描述。

表 4-4　　　　　　　　　Redis 发布和订阅的常用命令及其描述

常用命令	描述
PSUBSCRIBE	订阅一个或多个符合给定模式的频道
PUBSUB	查看发布和订阅系统状态
PUBLISH	将消息发送到指定的频道
PUNSUBSCRIBE	退订所有给定模式的频道
SUBSCRIBE	订阅给定的一个或多个频道的消息
UNSUBSCRIBE	只退订给定的频道

4.3.2 Redis 发布和订阅实例

以下实例演示了发布和订阅是如何工作的。

首先，打开一个客户端连接 Redis 服务器，作为订阅者接收消息。在本实例中我们创建了一个订阅频道，命名为 redisChat。

```
127.0.0.1:6379> SUBSCRIBE redisChat
Reading messages... (press Ctrl-C to quit)
1) "subscribe"
2) "redisChat"
3) (integer) 1
```

其次，重新开启一个 Redis 客户端，作为发送者发送消息，然后在同一个频道 redisChat 发布两次消息，订阅者就能接收消息。

```
127.0.0.1:6379> PUBLISH redisChat "message1"
(integer) 1
127.0.0.1:6379> PUBLISH redisChat "message2"
(integer) 1
127.0.0.1:6379>
```

订阅者的客户端会显示如下消息。

```
127.0.0.1:6379> SUBSCRIBE redisChat
Reading messages... (press Ctrl-C to quit)
1) "subscribe"
2) "redisChat"
3) (integer) 1
1) "message"
2) "redisChat"
3) "message1"
1) "message"
2) "redisChat"
3) "message2"
```

4.4 Redis 管道

Redis 是一个客户端-服务器（CS）模型的 TCP 服务器，使用和 HTTP 类似的相应请求协议。一个客户端可以通过一个 Socket 连续发送多个请求命令，每个请求命令发出后客户端通常会阻塞并等待 Redis 服务器处理，Redis 服务器处理完请求后会将结果通过响应报文返回给客户端。

Redis 的管道（Pipeline）可以一次性发送多条命令并在执行完后一次性将结果返回。管道可以减少客户端与 Redis 服务器的通信次数，从而降低往返延时时间。管道实现的原理是队列，而队列遵循先进先出原则，这样就保证了数据的顺序性。

Redis 管道技术可以在 Redis 服务器启动时使用，在 Redis 客户端向 Redis 服务器发送请求（Request），并一次性读取所有 Redis 服务器的响应（Response）。

在 Linux 客户端新建脚本 pipeline.sh。

```
[root@localhost ~]# touch pipeline.sh
```

使用 vi 命令修改 pipeline.sh。

```
[root@localhost ~]# vi pipeline.sh
```

在 pipeline.sh 中添加以下内容，脚本文件名为 Redis\Chapter04\pipeline.sh。

```
(echo -en "PING\r\n SET db redis\r\nGET db\r\nSET visitor 0\r\nINCR visitor\r\nINCR visitor\r\nINCR visitor\r\nGET visitor\r\n"; sleep 10) | nc localhost 6379
```

给 pipeline.sh 赋予可执行命令权限。

```
[root@localhost ~]# chmod +x pipeline.sh
```

执行 pipeline.sh。
```
[root@localhost ~]# ./pipeline.sh
```
得到以下内容。
```
[root@localhost ~]# ./pipeline.sh
+PONG
+OK
$5
redis
+OK
:1
:2
:3
$1
3
```
以上实例中,我们首先使用 PING 命令查看 Redis 服务器是否可用,之后我们设置了 db 的值为 redis 并获取了 db 的值,最后使 visitor 自增 3 次并获取了 visitor 的值。

打开一个终端,并使用 redis-cli 命令连接 Redis 服务器。
```
127.0.0.1:6379> GET visitor
"3"
127.0.0.1:6379> GET db
"redis"
```
从上面结果可以看出,Redis 数据库已经存储了 db 和 visitor 的值。

4.5 数据备份与恢复

1. 数据备份

Redis 的 SAVE 命令用于创建当前数据库的快照文件。

实例:使用 SAVE 命令创建当前数据库的快照。
```
127.0.0.1:6379> SAVE
OK
```
该命令将在 Redis 安装目录中创建 dump.rdb 文件。

2. 数据恢复

如果需要恢复数据,只需将一台 Redis 服务器上的快照文件(dump.rdb)移动到另一台 Redis 服务器的安装目录并启动服务即可。获取 Redis 安装目录可以使用 CONFIG 命令,如下所示。
```
127.0.0.1:6379> CONFIG GET dir
1) "dir"
2) "/usr/local/redis/bin"
```
在本例中使用 CONFIG GET dir 命令输出的 Redis 安装目录为/usr/local/redis/bin。

创建 Redis 备份文件也可以使用 BGSAVE 命令,该命令在后台执行。
```
127.0.0.1:6379> BGSAVE
Background saving started
```

4.6 Redis 性能测试

Redis 性能测试是通过同时执行多个命令实现的。Redis 性能测试的基本命令如下。
`redis-benchmark [option] [option value]`
实例：同时执行 10 000 个请求来测试 Redis 的性能，如图 4-6 所示。

```
[root@localhost ~]# redis-benchmark -n 10000
====== PING_INLINE ======
  10000 requests completed in 0.06 seconds
  50 parallel clients
  3 bytes payload
  keep alive: 1

100.00% <= 0 milliseconds
156250.00 requests per second

====== PING_BULK ======
  10000 requests completed in 0.08 seconds
  50 parallel clients
  3 bytes payload
  keep alive: 1

100.00% <= 1 milliseconds
128205.12 requests per second

====== SET ======
  10000 requests completed in 0.11 seconds
  50 parallel clients
  3 bytes payload
  keep alive: 1

99.30% <= 1 milliseconds
99.51% <= 4 milliseconds
99.64% <= 5 milliseconds
100.00% <= 5 milliseconds
92592.59 requests per second

====== GET ======
  10000 requests completed in 0.10 seconds
  50 parallel clients
  3 bytes payload
  keep alive: 1

98.23% <= 1 milliseconds
100.00% <= 1 milliseconds
101010.10 requests per second
```

图 4-6   redis-benchmark 命令同时执行 10 000 个请求

Redis 性能测试工具可选参数如表 4-5 所示。

表 4-5   Redis 性能测试工具可选参数

可选参数	描述	默认值
-h	指定服务器主机名	127.0.0.1
-p	指定服务器端口	6379
-s	指定服务器 Socket	
-c	指定并发连接数	50
-n	指定请求数	10 000
-d	以字节的形式指定 SET/GET 值的数据大小	2
-k	1=keep alive 0=reconnect	1
-r	SET/GET/INCR 使用随机 key，SADD 使用随机值	
-P	通过管道传输<numreq>请求	1
-q	强制退出 Redis。仅显示 query/sec 值	
--csv	以 CSV 格式输出	

续表

可选参数	描述	默认值
-l	生成循环，永久执行测试	
-t	仅运行以逗号分隔的测试命令列表	
-i	idle 模式，仅打开 N 个 idle 连接并等待	

实例：使用多个参数来测试 Redis 性能。

```
$ redis-benchmark -h 127.0.0.1 -p 6379 -t SET,LPSH -n 10000 -q
SET: 65789.48 requests per second
LPUSH: 74074.07 requests per second
```

以上实例中连接 Redis 的主机 IP 地址为 127.0.0.1，端口为 6379，执行 SET 命令和 LPUSH 命令，请求数为 10000，通过-q 参数让结果只显示每秒执行的请求数。

4.7　Redis 客户端连接

在 Redis 2.4 中，最大连接数是被直接硬编码写在代码里面的，而在 Redis 2.6 以后的版本中这个值变成可配置的。maxclients 的默认值是 10 000，也可以在 redis.conf 配置文件中对这个值进行修改。

```
127.0.0.1:6379> CONFIG GET maxclients
1) "maxclients"
2) "10000"
```

以下实例可以在 Redis 服务器启动时，设置最大连接数为 100 000。

```
$ redis-server --maxclients 100000
```

Redis 客户端查看客户端连接的命令及其描述如表 4-6 所示。

表 4-6　　　　　　Redis 客户端查看客户端连接的命令及其描述

命令	描述
CLIENT LIST	返回连接到 Redis 服务器的客户端列表
CLIENT SETNAME	设置当前连接的名称
CLIENT GETNAME	获取通过 CLIENT SETNAME 命令设置的服务名称
CLIENT PAUSE	挂起客户端连接，指定挂起的时间以毫秒计
CLIENT KILL	关闭客户端连接

Redis 客户端命令如图 4-7 所示。

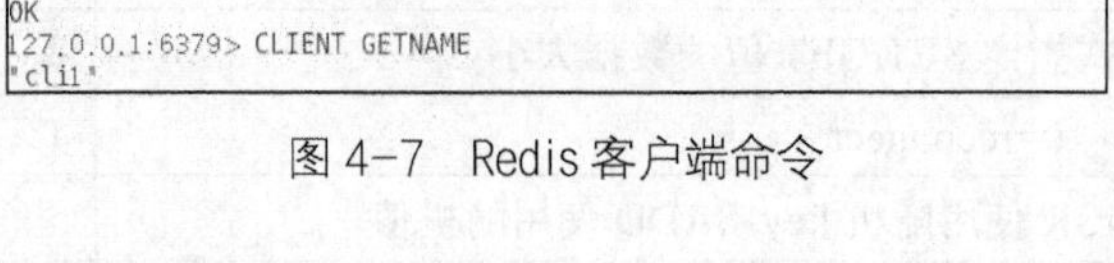

```
127.0.0.1:6379> CLIENT LIST
id=3703 addr=127.0.0.1:34430 fd=7 name= age=130 idle=0 flags=N db=0 sub
=0 psub=0 multi=-1 qbuf=26 qbuf-free=32742 obl=0 oll=0 omem=0 events=r
cmd=client
127.0.0.1:6379> CLIENT SETNAME cli1
OK
127.0.0.1:6379> CLIENT GETNAME
"cli1"
```

图 4-7　Redis 客户端命令

4.8　Redis 服务开机自启动

每次启动 Redis 服务都需要使用 redis-server 命令，稍显烦琐。当然也可以把 redis-server

命令放在启动脚本里,每次运行脚本就可以启动 Redis 服务了。但有没有更好的方法,让主机每次开机后就自动启动 Redis 服务呢?

答案是肯定的,这就是本节主要讨论的内容。

4.8.1 Windows 下 Redis 服务开机自启动

可以把 Redis 设置成 Windows 服务,这样 Windows 启动后就会自启动 Redis 服务,和 Windows 服务一样,可以启动/停止服务。

1. 注册 Redis 为 Windows 后台服务

在 Redis 的目录下执行以下命令,执行后 Redis 就作为 Windows 后台服务了。

```
redis-server --service-install redis.windows.conf
```

如果执行命令成功,会显示图 4-8 所示页面。

图 4-8 注册 Redis 为 Windows 后台服务

按"Win + R"组合键执行 services.msc 命令,就可以看到 Redis 已经作为 Windows 后台服务了。

2. 启动 Redis 服务

将 Redis 成功注册到 Windows 后台服务中后,Redis 并没有启动,可以在 Windows 服务列表中启动 Redis 服务。启动 Redis 服务的命令如下。

```
redis-server --service-start
```

3. 停止 Redis 服务

停止 Redis 服务的命令如下。

```
redis-server --service-stop
```

4. 卸载 Redis 服务

卸载 Redis 服务的命令如下。

```
redis-server --service-uninstall
```

将 Redis 成功注册到 Windows 后台服务后,就可以使用 Windows 命令启动/停止 Redis 服务,如图 4-9 所示。

图 4-9 使用 Windows 命令启动/停止 Redis 服务

启动 Redis 服务。
```
net start redis
```
停止 Redis 服务。
```
net stop redis
```

4.8.2 Linux 下 Redis 服务开机自启动

1. 修改 redis.conf 配置文件

为了让 redis-server 命令能在操作系统启动时自动执行，需要将 Redis 服务作为守护进程（Daemon）来运行。我们回到/usr/local/redis/conf 目录中找到 redis.conf 配置文件，这个文件是 Redis 服务运行时加载的配置文件，使用以下命令查看其中的内容。

```
$ vi /usr/local/redis/conf/redis.conf
```

此文件内容非常多，但是大部分内容是注释，我们重点关注其中的两个配置项：daemonize 和 pidfile。

- daemonize 默认值是 false，表示 Redis 服务作为守护进程来运行，需要把它改成 daemonize yes。
- pidfile 默认值是 pidfile /var/run/redis_6379.pid，表示当 Redis 服务以守护进程方式运行时，Redis 服务默认会把 pid 写入/var/run/redis_6379.pid 文件，Redis 服务运行时该文件就存在，Redis 服务一旦停止该文件就会自动删除，因而可以用来判断 Redis 服务是否正在运行。该配置项不用修改。

为了让 Redis 能在 Linux 启动时自动运行，需要修改完配置项 daemonize 后保存 redis.conf 配置文件，然后退出文件。

2. 修改初始化脚本 redis_init_script

有了基本配置，Redis 还需要有一个管理启动、关闭和重启的脚本。在 Redis 的源代码里已经提供了一个初始化脚本 redis_init_script。这个初始化脚本的位置在%/redis-6.0.6/utils/目录下，如图 4-10 所示。

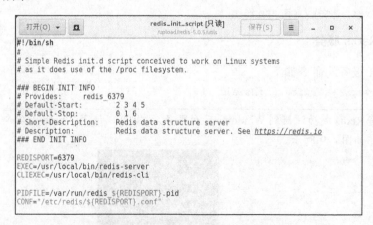

图 4-10 redis_init_script 脚本

redis_init_script 脚本中指定了端口、Server 路径、CLI 路径、PIDFILE 路径以及 CONF

路径。在安装时执行了 make install 命令后，这个脚本不需要做多大改动，因为 make install 命令会把 Redis 的可执行命令都复制到/usr/local/bin 目录下。

只需要修改 CONF 选项对应的 Redis 配置文件为 Linux 使用的 redis.conf 配置文件即可。在笔者的计算机上，Redis 配置文件保存在/usr/local/redis/conf/redis.conf 目录下，读者需要根据实际情况进行修改。修改后的 redis_init_script 脚本如图 4-11 所示。

图 4-11　修改后的 redis_init_script 脚本

3．将 redis_init_script 脚本复制到/etc/init.d 目录下并修改脚本名字为 redis

```
$ cp redis_init_script /etc/init.d/redis
```
给 redis 文件授予执行权限。
```
$ chmod +x /etc/init.d/redis
```

4．开启服务自启动

在/etc/init.d 目录下的脚本都是可以在 Linux 启动时自动启动的服务，还需要一个 Linux 启动时的配置。开启 Redis 服务自启动的命令如下。
```
$ chkconfig redis on
```

5．启动和停止 Redis 服务

重启 CentOS 之后，就可以执行以下命令启动和停止 Redis 服务了。

启动 Redis 服务。
```
$ service redis start
```
停止 Redis 服务。
```
$ service redis stop
```
以上两个命令等价于以下命令。
```
$/etc/init.d/redis start
$/etc/init.d/redis stop
```

4.9　Redis 内存分析工具

redis-rdb-tools 是用 Python 编写的用来分析 Redis 的 RDB 快照文件的工具。在内存分析中，我们主要用它生成内存报告。

1. 安装 redis-rdb-tools

安装 redis-rdb-tools 前需要配置 Python 环境，这部分的内容请参考 11.7.2 小节在 Linux 下安装 Python 3。配置好 Python 环境后可以通过 Python 的 pip 来安装 redis-rdb-tools。

```
$ pip3 install rdbtools python-lzf
```

2. 生成内存报告

使用 redis-rdb-tools 生成内存报告的实例如下。

首先，在 Redis 客户端生成两条 String 类型的数据，如下所示。

```
127.0.0.1:6379> SET name xinping
OK
127.0.0.1:6379> SET age 25
OK
127.0.0.1:6379> SAVE
```

然后，把 Redis 的 RDB 快照内存报告输出到控制台。

```
$ rdb -c memory /usr/local/redis/bin/dump.rdb
database,type,key,size_in_bytes,encoding,num_elements,len_largest_element,expiry
0,string,age,48,string,8,8,
0,string,name,64,string,7,7,
```

在生成的报告中有 database（key 所在的 Redis 数据库编号）、type（key 类型）、key（键）、size_in_bytes（key 占用的内存大小）、encoding（RDB 编码方式）、num_elements（key 中的 value 的个数）、len_largest_element（key 中的 value 的长度）和 expiry（key 的过期时间）

从返回的报告可以看出 key 为 age 的 String 类型数据占用了 46 个字节，key 为 name 的 String 类型数据占用了 64 个字节。

也可以使用以下命令把 Redis 的内存报告生成为 memory.csv 文件，并将文件输出到控制台。

```
$ rdb -c memory /usr/local/redis/bin/dump.rdb > memory.csv
$ cat memory.csv
database,type,key,size_in_bytes,encoding,num_elements,len_largest_element,expiry
0,string,age,48,string,8,8,
0,string,name,64,string,7,7,
```

第 5 章 Redis 缓存的持久化

在本章中，我们将学习 Redis 缓存持久化以及 Redis 过期 key 清除策略。

5.1 持久化机制

Redis 的所有数据都保存在内存中，如果没有配置持久化功能，Redis 重启后数据就会全部丢失，所以需要开启 Redis 的持久化功能，将数据保存到磁盘上，这样当 Redis 重启后，可以从磁盘中恢复数据。Redis 提供两种方式进行持久化，一种是 RDB 持久化，另一种是 AOF 持久化。下面详细介绍这两种方式。

1. RDB 持久化

RDB 持久化是指在指定的时间间隔内定时地将内存中的数据写入磁盘，把内存中的数据保存到 RDB 文件中，是默认的持久化方式。Redis 快照的过程是，首先 Redis 服务器使用 fork 函数复制一份当前进程（父进程）的副本（子进程）。然后，父进程继续接收并处理客户端发来的命令，而子进程将内存中的数据写入硬盘中的 RDB 临时文件。最后，当子进程写入完所有数据后会用 RDB 临时文件替换旧的 RDB 文件，如图 5-1 所示。

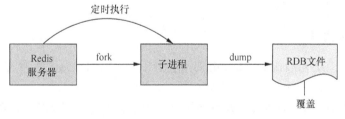

图 5-1 RDB 持久化

2. AOF 持久化

追加（Append Only File，AOF）持久化方式会记录 Redis 客户端对服务器的每一次写操作命令，并将这些写操作追加保存到 appendonly.aof 文件（AOF 文件）中。在 Redis 服务器重启时，会加载并运行 AOF 文件里的命令，以达到恢复数据的目的，如图 5-2 所示。

两种持久化是可以同时存在的，但是当 Redis 服务器重启时，AOF 文件会被优先用于重建数据。可以通过配置的方式禁用 Redis 服务器的持久化功能，让数据的生命周期只存在于

Redis 服务器的运行时间里，这样就可以将 Redis 服务器视为一个功能加强版的 Memcached。Memcached 是一个分布式的高速缓存系统，也是基于内存的 key-value 存储系统。通过 Redis 服务器作为缓存数据库进行查询，可以减少访问数据库的次数，以提高 Web 应用的访问速度。

图 5-2 AOF 持久化

5.1.1 配置 RDB

Redis 的配置文件在 Linux 下是 redis.conf 文件，在 Windows 下是 redis.windows.conf 文件。本小节使用的实验环境是 Linux，相应的配置文件是 redis.conf 文件，配置效果与 Windows 下的一致。

1．RDB 文件路径和名称

RDB 持久化是 Redis 默认的持久化方式，默认情况下 Redis 会把快照文件存储在当前目录下一个名为 dump.rdb 的文件内。

如果需要修改 dump.rdb 文件的存储路径和名称，可以通过修改配置文件 redis.conf 内的 dbfilename 参数和 dir 参数来实现。

```
# RDB 文件名，默认为 dump.rdb
dbfilename dump.rdb
```

```
# RDB 文件和 AOF 文件存放的目录。默认为当前工作目录
dir /usr/local/redis/bin
```

保存配置文件后，使用 redis-server 命令加载 redis.conf 配置文件并重新启动 Redis 服务器，然后使用 CONFIG GET dir 命令查看 RDB 文件的存储路径。

```
127.0.0.1:6379> CONFIG GET dir
1) "dir"
2) "/usr/local/redis/bin"
```

2．RDB 的保存点

修改保存点选项，可以配置 RDB 的启用和禁用。可以通过修改配置文件 redis.conf 实现该功能。

（1）设置保存点，可以使 Redis 在每 N 秒内，如果数据发生了 M 次改变就保存快照文件。例如，下面这个保存点配置表示每 60s 内，如果数据发生了 10000 次以上的改变，Redis 就会自动保存快照文件。

```
save 60 10000
```

保存点可以设置多个，设置保存点的格式如下。

```
save <seconds> <changes>
```

Redis 可以设置多个保存点，例如 Redis 的配置文件 redis.conf 就默认设置了 3 个保存点。

```
save 900 1        # 900s 后至少 1 个 key 有改变，就保存快照文件
save 300 10       # 300s 后至少 10 个 key 有改变，就保存快照文件
save 60 10000     # 60s 后至少 10000 个 key 有改变，就保存快照文件
```

(2) 禁用快照保存。如果想禁用快照保存的功能，可以通过注释所有 save 配置，或者在最后一条 save 配置后添加如下的配置实现。

```
save ""
```

3．错误处理

后台存储发生错误时禁止写入，默认值为 yes。默认情况下，如果 Redis 在后台生成快照文件时失败，就会停止接收数据，目的是让用户能知道数据没有持久化成功。

```
stop-writes-on-bgsave-error yes
```

4．数据压缩

启动 RDB 文件压缩，会耗费 CPU 资源，默认值为 yes。如果想节省 CPU 资源，可以禁用压缩功能，但是数据集就会比没压缩的时候要大。

```
rdbcompression yes
```

5．数据校验

对 RDB 数据进行校验，会耗费 CPU 资源，默认值为 yes。

```
rdbchecksum yes
```

6．手动生成快照文件

Redis 提供了 SAVE 命令和 BGSAVE 命令用于手动生成快照文件。

（1）SAVE。

SAVE 命令会使用同步的方式生成 RDB 快照文件，将当前 Redis 实例的所有数据快照（Snap Shot）以 RDB 文件的形式保存到硬盘，默认情况下会把 Redis 数据持久化到 dump.rdb 文件中，并且在 Redis 服务器重启后自动读取 dump.rdb 文件。SAVE 命令在 Redis 主线程中工作，会阻塞其他请求操作，在实际的生产环境中应该避免使用。

```
127.0.0.1:6379> SAVE
OK
```

（2）BGSAVE。

BGSAVE 命令使用异步的方式保存当前 Redis 实例的所有数据快照到 RDB 文件，执行 BGSAVE 命令后，Redis 会产生一个子进程并立刻恢复对客户端的服务。

```
127.0.0.1:6379> BGSAVE
Background saving started
(2.41s)
```

> Redis 配置文件里禁用了快照生成功能不影响 SAVE 命令和 BGSAVE 命令的效果。

5.1.2 配置 AOF

1．启用 AOF 持久化

将 redis.conf 配置文件的配置项 appendonly 设置为 yes，启用 AOF 持久化。

```
appendonly yes
```

修改 redis.conf 配置文件后，重启 Redis 服务器，Redis 执行的每一条命令都会被记录到 appendonly.aof 文件中。但事实上，Redis 并不会立即将命令写入硬盘文件中，而是将命令写入硬盘缓存。在接下来的可靠性配置中，可以配置从硬盘缓存写入硬盘的时间。

2. AOF 文件路径和名称

通过修改配置文件 redis.conf，修改 dir、appendfilename 对应的配置项来修改 AOF 文件路径和名称。

```
# 文件存放目录，与 RDB 文件共用。默认为当前工作目录
dir /usr/local/redis

# 默认文件名为 appendonly.aof
appendfilename "appendonly.aof"
```

3. 可靠性

在 redis.conf 配置文件中可以通过 appendfsync 选项指定写入策略，有 3 个选项。

- always：每次收到 Redis 客户端的写命令就立即强制写入到 AOF 文件。这是最有保证的持久化方式，但也是速度最慢的，一般不推荐使用。
- everysec：每秒向 AOF 文件写入一次 Redis 客户端的写操作。在性能和持久化方面做了很好的折中，是推荐的方式。
- no：由操作系统来决定什么时候写入 AOF 文件，一般为 30 秒左右一次。这个方式性能最好，但是持久化方面没有保证，一般不推荐使用。

在 redis.conf 配置文件中配置 appendfsync 选项的实例如下。

```
# appendfsync always
appendfsync everysec
# appendfsync no
```

4. 日志重写

随着写操作不断增加，AOF 文件会越来越大，Redis 可以在不中断服务的情况下在后台重建 AOF 文件。

日志重写的工作原理如下。

- Redis 调用 fork 函数，产生一个子进程。
- 子进程把新的 AOF 文件写到一个临时文件里。
- 父进程持续把新的变动写到内存里的缓冲区（buffer），同时也会把这些新的变动写到旧的 AOF 文件里，这样即使重写失败也能保证数据的安全。
- 当子进程完成文件的重写后，父进程会获得一个信号，然后把内存里的缓冲区内容追加到子进程生成的新 AOF 文件里。

我们可以设置日志重写的条件。下面的实例表示当 AOF 文件的体积大于 64 MB，并且 AOF 文件的体积比上一次重写之后的体积大了至少一倍（100%）的时候，Redis 将执行日志重写操作。

```
auto-aof-rewrite-percentage 100
```

```
auto-aof-rewrite-min-size 64MB
```
Redis 会记住自从上一次重写后 AOF 文件的大小。要禁用自动的日志重写功能,可以把百分比设置为 0。
```
auto-aof-rewrite-percentage 0
```

5.数据损坏修复

如果因为某些原因(例如服务器崩溃)AOF 文件损坏了,导致 Redis 加载不了,可以通过以下方式进行修复。

- 备份 AOF 文件。
- 使用 redis-check-aof 命令修复原始的 AOF 文件。
```
redis-check-aof --fix
```
- 在 Linux 下可以使用 diff -u 命令查看 AOF 文件和 RDB 文件的差异。
- 使用修复过的 AOF 文件重启 Redis 服务器。

6.从 RDB 切换到 AOF

在 Redis 2.2 以后的版本中,从 RDB 切换到 AOF,需要备份一个最新的 dump.rdb 文件,并把备份文件放在一个安全的地方。执行以下命令。
```
$ redis-cli config set appendonly yes
$ redis-cli config set save ""
```
要确保数据与切换前一致,确保数据正确地写到 AOF 文件里。

7.备份

建议的备份方法如下。

- 创建一个定时任务,每小时和每天创建一个快照文件,并将快照文件保存在不同的文件夹里。
- 定时任务运行时,把太旧的文件删除。例如只保留 48h 内的按小时创建的快照文件和一到两个月的按天创建的快照文件。
- 每天确保一次把快照文件传输到数据中心外的地方进行保存,至少不能保存在 Redis 服务所在的服务器。

5.2 Redis 过期 key 清除策略

Redis 对过期 key 有 3 种清除策略。

(1)被动清除:当读/写一个已经过期的 key 时,会触发惰性清除策略,直接清除这个过期 key。

(2)主动清除:由于惰性清除策略无法保证冷数据被及时清除,因此 Redis 会定期主动清除一批已过期的 key。

(3)当前已用内存超过 maxmemory 限定时,触发主动清除策略。

这里着重介绍第 3 种清除策略。在 Redis 中,允许用户设置最大使用内存大小为 maxmemory(需要配合 maxmemory-policy 使用),设置为 0 表示不限制。当 Redis 内存数据集快到达 maxmemory 时，Redis 会实行数据淘汰策略。Redis 提供 6 种数据淘汰策略，如表 5-1 所示。

表 5-1　　　　　　　　　Redis 提供的 6 种数据淘汰策略及其描述

数据淘汰策略	描述
volatile-lru	从已设置过期时间的数据集中，挑选最近最少使用的数据淘汰
volatile-ttl	从已设置过期时间的数据集中，挑选即将过期的数据淘汰
volatile-random	从已设置过期时间的数据集中，随机挑选数据淘汰
allkeys-lru	从所有的数据集中，挑选最近最少使用的数据淘汰
allkeys-random	从所有的数据集中，随机挑选数据淘汰
no-enviction	禁止淘汰数据，这是默认淘汰策略

关于 maxmemory 设置，可以通过在 redis.conf 配置文件中的 maxmemory 参数设置，或者通过命令 CONFIG SET 动态修改。关于数据淘汰策略的设置，可以通过在 redis.conf 配置文件中的 maxmemory-policy 参数设置，或者通过命令 CONFIG SET 动态修改。

```
127.0.0.1:6379> CONFIG GET maxmemory
1) "maxmemory"
2) "0"
127.0.0.1:6379> CONFIG SET maxmemory 100MB
OK
127.0.0.1:6379> CONFIG GET maxmemory
1) "maxmemory"
2) "104857600"
```

第 6 章 Redis 集群环境部署

前文介绍的 Redis，我们都是在一台服务器上进行操作的，也就是说读、写以及备份操作都是在一台 Redis 服务器上进行的。随着项目访问量的增加，对 Redis 服务器的操作也更加频繁，虽然 Redis 读写速度都很快，但是一定程度上也会造成一定的延时。为了解决访问量大的问题，通常会采取的一种方式是主从（Master-Slave）复制。Master 以写为主，Slave 以读为主，Master 更新后根据配置自动同步到 Slave。

本节中 Master 表示主节点，Slave 表示从节点，以后不再赘述。

本章主要介绍 Redis 集群环境部署，讲解主从复制、哨兵模式和 Redis 集群。

6.1 主从复制

主从复制也叫主从模式，当用户向 Master 写入数据时，Master 通过 Redis 同步机制将数据文件发送至 Slave，Slave 也会通过 Redis 同步机制将数据文件发送至 Master 以确保数据一致，从而实现 Redis 的主从复制。如果 Master 和 Slave 之间的连接中断，Slave 可以自动重连 Master，但是连接成功后，将自动执行一次完全同步。

配置主从复制后，Master 可以负责读写服务，Slave 只负责读服务。Redis 复制在 Master 这一端是非阻塞的，也就是说在和 Slave 同步数据的时候，Master 仍然可以执行客户端的命令而不受其影响。

1．主从复制的特点

（1）同一个 Master 可以拥有多个 Slave。

（2）Master 下的 Slave 还可以接受同一架构中其他 Slave 的连接与同步请求，实现数据的级联复制，即 Master→Slave→Slave 模式。

（3）Master 以非阻塞的方式同步数据至 Slave，这将意味着 Master 会继续处理一个或多个 Slave 的读写请求。

（4）主从复制不会阻塞 Master，当一个或多个 Slave 与 Master 进行初次同步数据时，Master 可以继续处理客户端发来的请求。

（5）主从复制具有可扩展性，即多个 Slave 专门提供只读查询与数据的冗余，Master 专门提供写操作。

（6）通过配置禁用 Master 数据持久化机制，将其数据持久化操作交给 Slave 完成，避免在 Master 中有独立的进程来完成此操作。

2．主从复制的优势

- 避免 Redis 单点故障。
- 做到读写分离，构建读写分离架构，满足读多写少的应用场景。

6.1.1　Redis 主从复制原理

当启动一个 Slave 进程后，它会向 Master 发送一个 SYNC 命令，请求同步连接。无论是第一次连接还是重新连接，Master 都会启动一个后台进程，将数据快照保存到数据文件中，同时 Master 会记录所有修改数据的命令并将其缓存在数据文件中。

后台进程完成缓存操作后，Master 就发送数据文件给 Slave，Slave 将数据文件保存到硬盘上，然后将其加载到内存中。接着 Master 就会把所有修改数据的命令发送给 Slave。

若 Slave 出现故障导致宕机，那么恢复正常后会自动重新连接。Master 收到 Slave 的连接请求后，将其完整的数据文件发送给 Slave。如果 Mater 同时收到多个 Slave 发来的同步请求，那么 Master 只会在后台启动一个进程保存数据文件，然后将其发送给所有的 Slave，确保 Slave 正常。

关于 Redis 的安装与配置这里不再重复，具体安装过程请读者参考 1.2.2 节在 Linux 下安装 Redis。Redis 主从复制环境使用的服务器资源如表 6-1 所示，将 Redis 安装在 3 台服务器上，3 台服务器的操作系统都是 CentOS 7。

表 6-1　　　　　　　　　Redis 主从复制环境使用的服务器资源

端口	操作系统	IP 地址	角色
6379	CentOS 7 x86_64	192.168.11.10	Master
		192.168.11.11	Slave1
		192.168.11.12	Slave2

6.1.2　Redis 主从复制安装过程

1．Master 操作

在 Redis 主服务器上的 redis.conf 配置文件中修改 bind 字段，将以下内容：
```
bind 127.0.0.1
```
修改为 Master 的主机 IP 地址。
```
bind 127.0.0.1 192.168.11.10
```
如果 Redis 主服务器只绑定了 127.0.0.1，那么跨服务器 IP 地址的访问就会失败，也就是只有本机才能访问，外部请求会被过滤，这是由 Linux 的网络安全策略管理的。如果绑定的 IP 地址只是 192.168.11.10，那么本机通过 localhost 和 127.0.0.1，或者直接输入命令 redis-cli 登录本机 Redis 就会失败。所以跨服务器访问 Redis，需要加上服务器 IP 地址才能被访问。

运行 Redis 服务。
```
$ redis-server /usr/local/redis/conf/redis.conf
```

2. Slave1 操作

修改 Redis 的配置文件。

```
$ vi /usr/local/redis/conf/redis.conf
```

添加 Master 的 IP 地址与端口。

```
slaveof 192.168.11.10 6379
```

添加 Mater 的 IP 地址和端口时，中间用空格分隔，然后保存 redis.conf 配置文件。

运行 Slave1（192.168.11.11）的 Redis。

```
$ redis-server /usr/local/redis/conf/redis.conf
```

查看 Slave1 运行日志，如图 6-1 所示。

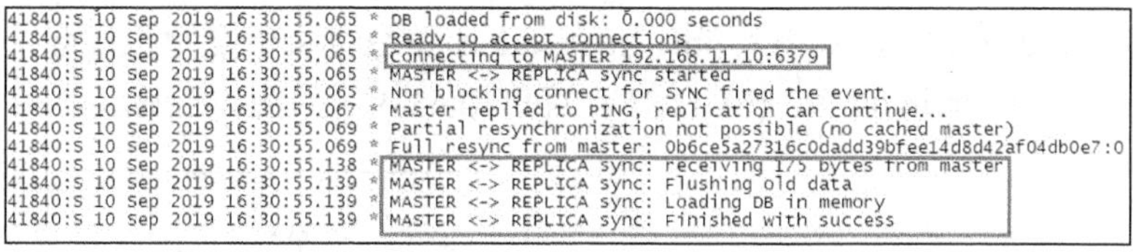

```
41840:S 10 Sep 2019 16:30:55.065 * DB loaded from disk: 0.000 seconds
41840:S 10 Sep 2019 16:30:55.065 * Ready to accept connections
41840:S 10 Sep 2019 16:30:55.065 * Connecting to MASTER 192.168.11.10:6379
41840:S 10 Sep 2019 16:30:55.065 * MASTER <-> REPLICA sync started
41840:S 10 Sep 2019 16:30:55.065 * Non blocking connect for SYNC fired the event.
41840:S 10 Sep 2019 16:30:55.067 * Master replied to PING, replication can continue...
41840:S 10 Sep 2019 16:30:55.069 * Partial resynchronization not possible (no cached master)
41840:S 10 Sep 2019 16:30:55.069 * Full resync from master: 0b6ce5a27316c0dadd39bfee14d8d42af04db0e7:0
41840:S 10 Sep 2019 16:30:55.138 * MASTER <-> REPLICA sync: receiving 175 bytes from master
41840:S 10 Sep 2019 16:30:55.139 * MASTER <-> REPLICA sync: Flushing old data
41840:S 10 Sep 2019 16:30:55.139 * MASTER <-> REPLICA sync: Loading DB in memory
41840:S 10 Sep 2019 16:30:55.139 * MASTER <-> REPLICA sync: Finished with success
```

图 6-1 Slave1 运行日志

查看 Master 运行日志，如图 6-2 所示。

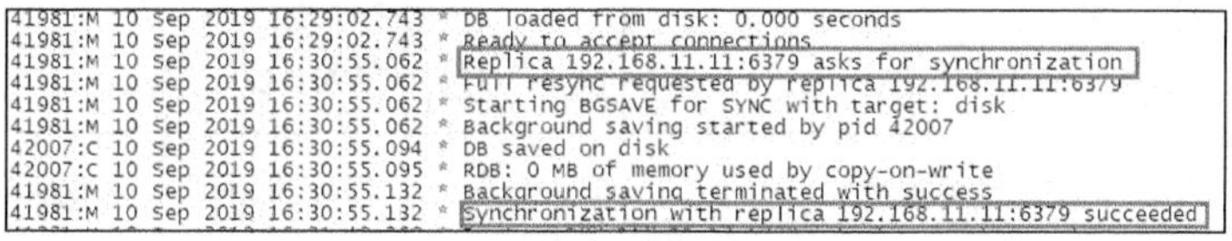

```
41981:M 10 Sep 2019 16:29:02.743 * DB loaded from disk: 0.000 seconds
41981:M 10 Sep 2019 16:29:02.743 * Ready to accept connections
41981:M 10 Sep 2019 16:30:55.062 * Replica 192.168.11.11:6379 asks for synchronization
41981:M 10 Sep 2019 16:30:55.062 * Full resync requested by replica 192.168.11.11:6379
41981:M 10 Sep 2019 16:30:55.062 * Starting BGSAVE for SYNC with target: disk
41981:M 10 Sep 2019 16:30:55.062 * Background saving started by pid 42007
42007:C 10 Sep 2019 16:30:55.094 * DB saved on disk
42007:C 10 Sep 2019 16:30:55.095 * RDB: 0 MB of memory used by copy-on-write
41981:M 10 Sep 2019 16:30:55.132 * Background saving terminated with success
41981:M 10 Sep 2019 16:30:55.132 * Synchronization with replica 192.168.11.11:6379 succeeded
```

图 6-2 Master 运行日志

3. Slave2 操作

修改 Redis 的配置文件，添加 Master 的 IP 地址与端口。

```
$ vi /usr/local/redis/conf/redis.conf
slaveof 192.168.1.10 6379
```

添加 Master 的 IP 地址和端口时，中间用空格分隔，然后保存 redis.conf 配置文件。

运行 Slave2（192.168.11.12）的 Redis。

```
$ redis-server /usr/local/redis/conf/redis.conf
```

查看 Slave2 运行日志，如图 6-3 所示。

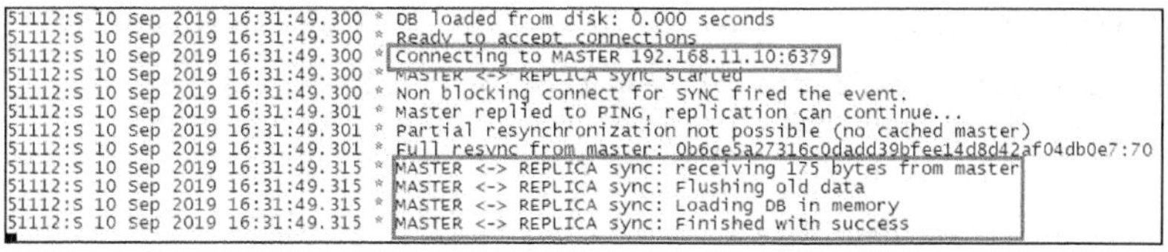

```
51112:S 10 Sep 2019 16:31:49.300 * DB loaded from disk: 0.000 seconds
51112:S 10 Sep 2019 16:31:49.300 * Ready to accept connections
51112:S 10 Sep 2019 16:31:49.300 * Connecting to MASTER 192.168.11.10:6379
51112:S 10 Sep 2019 16:31:49.300 * MASTER <-> REPLICA sync started
51112:S 10 Sep 2019 16:31:49.300 * Non blocking connect for SYNC fired the event.
51112:S 10 Sep 2019 16:31:49.301 * Master replied to PING, replication can continue...
51112:S 10 Sep 2019 16:31:49.301 * Partial resynchronization not possible (no cached master)
51112:S 10 Sep 2019 16:31:49.301 * Full resync from master: 0b6ce5a27316c0dadd39bfee14d8d42af04db0e7:70
51112:S 10 Sep 2019 16:31:49.315 * MASTER <-> REPLICA sync: receiving 175 bytes from master
51112:S 10 Sep 2019 16:31:49.315 * MASTER <-> REPLICA sync: Flushing old data
51112:S 10 Sep 2019 16:31:49.315 * MASTER <-> REPLICA sync: Loading DB in memory
51112:S 10 Sep 2019 16:31:49.315 * MASTER <-> REPLICA sync: Finished with success
```

图 6-3 Slave2 运行日志

结果与 Slave1 类似，只不过 Slave2 与 Slave1（192.168.11.11:6379）建立连接，在同步数据时，Redis 的主从级联复制便是这样：Master→Slave1→Slave2。

6.1.3 Redis 测试主从复制关系

1. 通过 info replication 命令查看节点角色

在主节点（192.168.11.10）输入 info replication 命令查看节点角色，会发现角色是 Master，有两个从节点 Slave1（192.168.11.11）和 Slave2（192.168.11.12），如图 6-4 所示。

在 info replication 命令的返回信息中“slave0:ip=192.168.11.11,port=6379,state=online,offset=1681,lag=0”，表示 slave0 是个从节点，使用的 IP 地址是 192.168.11.11，端口是 6379，状态是 online（在线状态）。这里的 slave0 对应上一节中配置的 Slave1 从节点，类似地，在返回信息中的 Slave1 对应上一节中配置的 Slave2 从节点。

在从节点（192.168.11.11）输入 info replication 命令查看节点角色，会发现节点角色是 Slave，它的主节点 IP 地址是 192.168.11.10，端口是 6379，如图 6-5 所示。

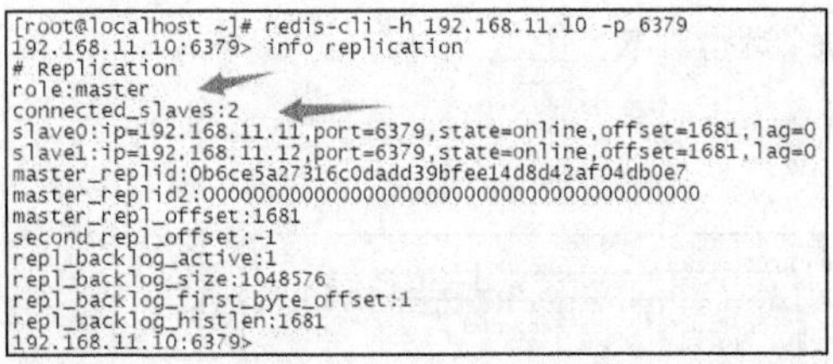

```
[root@localhost ~]# redis-cli -h 192.168.11.10 -p 6379
192.168.11.10:6379> info replication
# Replication
role:master
connected_slaves:2
slave0:ip=192.168.11.11,port=6379,state=online,offset=1681,lag=0
slave1:ip=192.168.11.12,port=6379,state=online,offset=1681,lag=0
master_replid:0b6ce5a27316c0dadd39bfee14d8d42af04db0e7
master_replid2:0000000000000000000000000000000000000000
master_repl_offset:1681
second_repl_offset:-1
repl_backlog_active:1
repl_backlog_size:1048576
repl_backlog_first_byte_offset:1
repl_backlog_histlen:1681
192.168.11.10:6379>
```

图 6-4　主节点角色

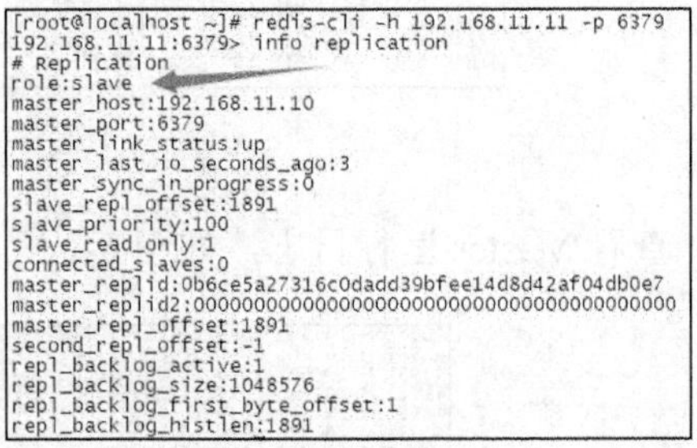

```
[root@localhost ~]# redis-cli -h 192.168.11.11 -p 6379
192.168.11.11:6379> info replication
# Replication
role:slave
master_host:192.168.11.10
master_port:6379
master_link_status:up
master_last_io_seconds_ago:3
master_sync_in_progress:0
slave_repl_offset:1891
slave_priority:100
slave_read_only:1
connected_slaves:0
master_replid:0b6ce5a27316c0dadd39bfee14d8d42af04db0e7
master_replid2:0000000000000000000000000000000000000000
master_repl_offset:1891
second_repl_offset:-1
repl_backlog_active:1
repl_backlog_size:1048576
repl_backlog_first_byte_offset:1
repl_backlog_histlen:1891
```

图 6-5　从节点角色 1

在从节点（192.168.11.12）输入 info replication 命令查看节点角色，也会发现节点角色是 Slave，它的主节点 IP 地址是 192.168.11.10，端口是 6379，如图 6-6 所示。

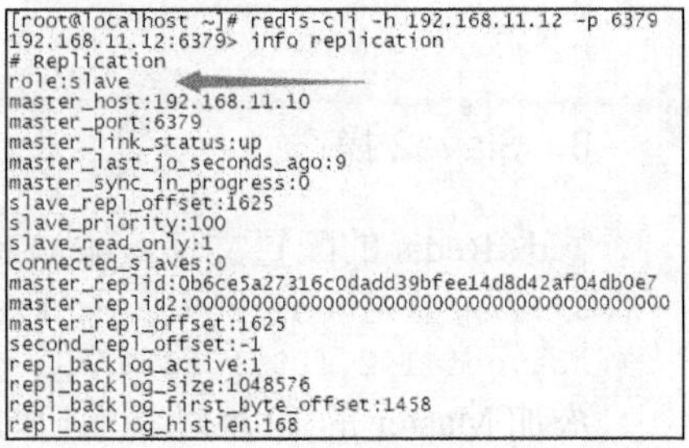

```
[root@localhost ~]# redis-cli -h 192.168.11.12 -p 6379
192.168.11.12:6379> info replication
# Replication
role:slave
master_host:192.168.11.10
master_port:6379
master_link_status:up
master_last_io_seconds_ago:9
master_sync_in_progress:0
slave_repl_offset:1625
slave_priority:100
slave_read_only:1
connected_slaves:0
master_replid:0b6ce5a27316c0dadd39bfee14d8d42af04db0e7
master_replid2:0000000000000000000000000000000000000000
master_repl_offset:1625
second_repl_offset:-1
repl_backlog_active:1
repl_backlog_size:1048576
repl_backlog_first_byte_offset:1458
repl_backlog_histlen:168
```

图 6-6　从节点角色 2

2. 测试主从读写分离

在 Master 操作如下，由客户端验证同步结果。

```
$ redis-cli -h 192.168.1.10 -p 6379
192.168.11.10:6379> SET name xinping
OK
192.168.11.10:6379>
```

在 Slave1 验证同步结果。

```
$ /usr/local/redis/bin/redis-cli -h 192.168.11.11 -p 6379
192.168.11.11:6379> KEYS *
1) "name"
192.168.11.11:6379> GET name
"xinping"
192.168.11.11:6379>
```

在 Slave2 验证同步结果。

```
$ /usr/local/redis/bin/redis-cli  -h 192.168.11.12 -p 6379
192.168.11.12:6379> KEYS *
```

```
1) "name"
192.168.11.12:6379> GET name
"xinping"
192.168.11.12:6379>
```

主从复制只能在主节点执行写操作，在从节点执行写操作会报异常“(error) READONLY You can't write against a read only replica.”，如图 6-7 所示。

```
[root@localhost ~]# redis-cli -h 192.168.11.12 -p 6379
192.168.11.12:6379> set age 20
(error) READONLY You can't write against a read only replica.
```

图 6-7 在从节点执行写操作报异常

3. 测试主节点宕机

本实例测试当主节点宕机时，两个从节点的角色是否会发生变化。

首先，在主节点使用 shutdown 命令，模拟计算机故障停止主节点的服务。

```
[root@localhost ~]# redis-cli -h 192.168.11.10 -p 6379
192.168.11.10:6379> shutdown
```

然后，在从节点使用 info replication 命令查看角色。从图 6-8 中可以看出，当主节点宕机后，从节点的角色没有发生变化，不过 master_link_status 的状态变为 down，意味着此时从节点执行读取命令，无法承担起主节点的任务，从节点只能执行读操作。

```
[root@localhost ~]# redis-cli -h 192.168.11.11 -p 6379
192.168.11.11:6379> info replication
# Replication
role:slave
master_host:192.168.11.10
master_port:6379
master_link_status:down
master_last_io_seconds_ago:-1
master_sync_in_progress:0
slave_repl_offset:0
master_link_down_since_seconds:62
slave_priority:100
slave_read_only:1
connected_slaves:0
master_replid:da074e9e7f60c5049e4cc64dae42c1f33347ec76
master_replid2:0000000000000000000000000000000000000000
master_repl_offset:0
second_repl_offset:-1
repl_backlog_active:1
repl_backlog_size:1048576
repl_backlog_first_byte_offset:1
repl_backlog_histlen:0
```

```
[root@localhost ~]# redis-cli -h 192.168.11.12 -p 6379
192.168.11.12:6379> info replication
# Replication
role:slave
master_host:192.168.11.10
master_port:6379
master_link_status:down
master_last_io_seconds_ago:-1
master_sync_in_progress:0
slave_repl_offset:0
master_link_down_since_seconds:99
slave_priority:100
slave_read_only:1
connected_slaves:0
master_replid:da074e9e7f60c5049e4cc64dae42c1f33347ec76
master_replid2:0000000000000000000000000000000000000000
master_repl_offset:0
second_repl_offset:-1
repl_backlog_active:1
repl_backlog_size:1048576
repl_backlog_first_byte_offset:1
repl_backlog_histlen:0
```

图 6-8 主节点宕机后，两个从节点的角色没有发生变化

通过主从配置的实例，我们可以看出主节点只有一个，一旦主节点宕机之后，从节点无法承担起主节点的任务，那么整个系统也无法运行。如果主节点宕机之后，从节点能够自动变成主节点，那么问题就解决了，于是哨兵模式诞生了，这就是我们 6.2 节主要讨论的内容。

6.2 哨兵模式

6.2.1 灾备切换 Sentinel 的使用

Redis 2.6 中开始提供了哨兵模式，到 Redis 2.8 以后的版本中该模式正式稳定。哨兵（Sentinel）进程监控 Redis 集群中 Master 主服务器工作的状态，在 Master 发生故障的时候，可以实现 Master 和 Slave 的切换，保证系统的高可用性。哨兵模式的出现是为了解决主从复制的缺点，其架构如图 6-9 所示。

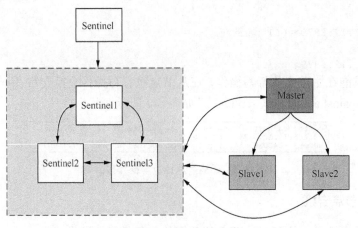

图 6-9　哨兵模式架构

1. 哨兵模式的基本概念

- Master 节点：主节点，Redis 的主数据库，写入都在这个节点上。
- Slave 节点：从节点，Redis 的从数据库，读取都在这个节点上。
- Sentinel 节点：哨兵节点，监控各个节点的状态。

基于哨兵模式的高可用架构如图 6-10 所示。

在这个架构中，复制主要是将主节点的数据同步到从节点，这样做主要有以下两个原因。

- 一旦主节点宕机了，从节点可以作为主节点的备份随时成为新的主节点。
- 从节点可以作为主节点分担读的压力。

2. 哨兵进程的作用

（1）监控（Monitoring）：哨兵进程会不断地检查 Master 和 Slave 是否运作正常。

（2）提醒（Notification）：当被监控的某个节点出现问题时，哨兵进程可以通过 API 向管理员或者其他应用程序发送通知。

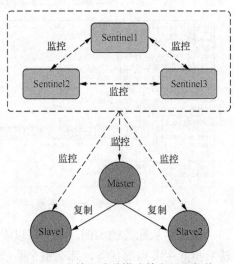

图 6-10　基于哨兵模式的高可用架构

（3）自动故障迁移（Automatic Failover）：当一个 Master 不能正常工作时，哨兵进程会开始一次自动故障迁移操作，它会将失效 Master 的其中一个 Slave 升级为新的 Master，并让失效 Master 的其他 Slave 改为复制新的 Master。当客户端试图连接失效的 Master 时，Redis 集群也会向客户端返回新 Master 的地址，使得 Redis 集群可以使用现在的 Master 替换失效 Master。Redis Sentinel 故障转移架构如图 6-11 所示。

（4）配置提供者：在哨兵模式下，客户端在初始化时连接的是哨兵节点集合，从中获取主节点的信息。

3. 部署技巧

- 在生产环境中 Sentinel 节点不应该部署在一台物理"计算机"上。

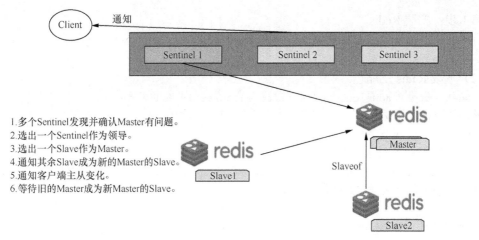

图 6-11　Redis Sentinel 故障转移架构

- 在生产环境中部署至少 3 个且奇数个 Sentinel 节点。

6.2.2　Redis Sentinel 的安装与配置

本小节我们配置一个（Master）和两个（Slave），并在一台服务器上部署 Redis 服务器和 Sentinel 实例。哨兵模式的实验环境如表 6-2 所示。

表 6-2　　　　　　　　　　　哨兵模式的实验环境

端口	操作系统	IP 地址	角色
7000	CentOS 7 x86_64	192.168.11.13	Master
7001			Slave1
7002			Slave2
26379			Sentinel1
26380			Sentinel2
26381			Sentinel3

Redis Sentinel 的主从架构如图 6-12 所示。

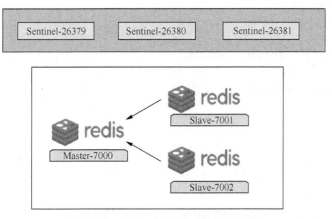

图 6-12　Redis Sentinel 的主从架构

第 1 步：安装 Redis。
（1）首先，安装 gcc 基础依赖包，使用如下命令。
```
$ yum install gcc-c++
```
（2）下载 Redis 压缩包。
```
$ wget http://download.redis.io/releases/redis-6.0.6.tar.gz
```
解压缩 Redis 压缩包。
```
$ tar -xzvf redis-6.0.6.tar.gz
```
进入解压缩后的目录。
```
$ cd redis-6.0.6/
```
使用 make 命令编译 Redis 源文件。
```
$ make
```
编译成功后，安装 Redis。
```
$ make install PREFIX=/usr/local/redis
```
安装成功后，需要对 Redis 进行部署，把 Redis 的配置文件 sentinel.conf 复制到 /usr/local/redis/conf 目录下。
```
$ mkdir /usr/local/redis/conf
$ cp sentinel.conf /usr/local/redis/conf
```
（3）配置 Redis 的命令。
最后需要将 Redis 的命令所在目录添加到系统变量 Path 中，修改 /etc/profile 文件。
```
$ vi /etc/profile
```
在 /etc/profile 文件最后一行添加以下内容。
```
export PATH=$PATH:/usr/local/redis/bin
```
使用 source 命令使 /etc/profile 文件生效。
```
$ source /etc/profile
```
至此，Redis 在 Linux 上的安装和配置就结束了。
第 2 步：主从节点的安装和配置。
在 /usr/local 下创建文件夹 redis-sential，使用如下命令。
```
$ mkdir -p /usr/local/redis-sential
$ cd /usr/local/redis-sential/
```
（1）安装主节点。
建立配置目录。
```
$ mkdir -p /opt/soft/redis/data/
$ mkdir -p /var/run/
```
创建主节点的配置文件 redis-7000.conf。
```
$ vi redis-7000.conf
```
在主节点的配置文件中添加以下内容。
```
port 7000
daemonize yes
pidfile /var/run/redis-7000.pid
logfile "7000.log"
dir "/opt/soft/redis/data/"
```
（2）创建从节点。
创建从节点的配置文件 redis-7001.conf。
```
$ vi redis-7001.conf
```

在从节点的配置文件中添加以下内容。
```
port 7001
daemonize yes
pidfile /var/run/redis-7001.pid
logfile "7001.log"
dir "/opt/soft/redis/data/"
slaveof 127.0.0.1 7000
```
创建从节点的配置文件 redis-7002.conf。
```
$ vi redis-7002.conf
```
在从节点的配置文件中添加以下内容。
```
port 7002
daemonize yes
pidfile /var/run/redis-7002.pid
logfile "7002.log"
dir "/opt/soft/redis/data/"
slaveof 127.0.0.1 7000
```
（3）快速启动节点。

启动主节点 7000。
```
$ redis-server /usr/local/redis-sential/redis-7000.conf
```
查看是否连接了主节点 7000。
```
[root@localhost redis-sential]# redis-cli -p 7000
127.0.0.1:7000> PING
PONG
```
启动两个从节点 7001、7002。
```
$ redis-server /usr/local/redis-sential/redis-7001.conf
$ redis-server /usr/local/redis-sential/redis-7002.conf
```
查看 Redis 进程。
```
[root@localhost ~]# ps -ef | grep redis
root      4356     1  0 16:56 ?        00:00:00 redis-server *:7000
root      4582     1  0 17:09 ?        00:00:00 redis-server *:7001
root      4596     1  0 17:09 ?        00:00:00 redis-server *:7002
root      4615  4480  0 17:10 pts/1    00:00:00 grep --color=auto redis
```
查看主从复制的关系。
```
[root@localhost redis-sential]# redis-cli -p 7000
127.0.0.1:7000> info replication
# Replication
role:master
connected_slaves:2
slave0:ip=127.0.0.1,port=7001,state=online,offset=168,lag=1
slave1:ip=127.0.0.1,port=7002,state=online,offset=168,lag=0
master_replid:85839c786b670a193462c2f77d043032b71a0c5c
master_replid2:0000000000000000000000000000000000000000
master_repl_offset:168
second_repl_offset:-1
repl_backlog_active:1
repl_backlog_size:1048576
repl_backlog_first_byte_offset:1
repl_backlog_histlen:168
```

从以上信息可以看出主节点 7000 的角色是 Master，有两个从节点 Slave0 和 Slave1，它们占用的端口分别是 7001 和 7002。

第 3 步：配置开启 Sentinel 监控主节点。

Sentinel 是特殊的 Redis，Sentinel 主要负责监控故障转移和通知。Sentinel 默认端口是 26379。复制 sentinel.conf 配置文件到指定目录。

```
$ cp /usr/local/redis/conf/sentinel.conf /usr/local/redis-sential
```

将 sentinel.conf 配置文件所有的空行和注释去掉。

```
$ cat sentinel.conf | grep -v "#" | grep -v "^$" > redis-sentinel-26379.conf
```

查看 sentinel.conf 配置文件发现所有的空行和注释都去掉了。

```
[root@localhost redis-sential]# cat redis-sentinel-26379.conf
port 26379
daemonize no
pidfile /var/run/redis-sentinel.pid
logfile ""
dir /tmp
sentinel monitor mymaster 127.0.0.1 6379 2
sentinel down-after-milliseconds mymaster 30000
sentinel parallel-syncs mymaster 1
sentinel failover-timeout mymaster 180000
sentinel deny-scripts-reconfig yes
```

然后再对 redis-sentinel-26379.conf 配置文件进行修改。

```
$ vi redis-sentinel-26379.conf
```

添加以下内容。

```
port 26379
daemonize yes
pidfile /var/run/redis-sentinel.pid
logfile "26379.log"
dir /opt/soft/redis/data
sentinel monitor mymaster 127.0.0.1 7000 2
sentinel down-after-milliseconds mymaster 30000
sentinel parallel-syncs mymaster 1
sentinel failover-timeout mymaster 180000
sentinel deny-scripts-reconfig yes
```

使用以下命令启动 Sentinel。

```
$ redis-sentinel redis-sentinel-26379.conf
```

查看 Sentinel 进程，可以看出已经启动了。

```
[root@localhost redis-sential]# ps -ef | grep redis-sentinel
root       7045      1  0 20:32 ?        00:00:03 redis-sentinel *:26379 [sentinel]
root       7440   2629  0 21:08 pts/0    00:00:00 grep --color=auto redis-sentinel
```

使用 redis-cli 命令连接到 Sentinel 节点。

```
$ redis-cli -p 26379
```

连接成功。

```
[root@localhost redis-sential]# redis-cli -p 26379
127.0.0.1:26379> ping
PONG
```

输入 info sentinel 命令会返回如下信息。

```
127.0.0.1:26379> info sentinel
```

```
# Sentinel
sentinel_masters:1
sentinel_tilt:0
sentinel_running_scripts:0
sentinel_scripts_queue_length:0
sentinel_simulate_failure_flags:0
master0:name=mymaster,status=ok,address=127.0.0.1:7000,slaves=2,sentinels=1
```
可以看到有一个 Master 和两个 Slave，只有一个 Sentinel。

还是查看 redis-sentinel-26379.conf 配置文件，在文件的最下面会发现多了 4 行内容，这是配置重写操作时产生的内容。

```
[root@localhost redis-sential]# cat redis-sentinel-26379.conf
port 26379
daemonize yes
pidfile "/var/run/redis-sentinel.pid"
logfile "26379.log"
dir "/opt/soft/redis/data"
sentinel myid 6c6fb93d92461803a7df349af5f1ba280c1c0a41
sentinel deny-scripts-reconfig yes
sentinel monitor mymaster 127.0.0.1 7000 2
sentinel config-epoch mymaster 0
sentinel leader-epoch mymaster 0
# Generated by CONFIG REWRITE
protected-mode no
sentinel known-replica mymaster 127.0.0.1 7002
sentinel known-replica mymaster 127.0.0.1 7001
sentinel current-epoch 0
```
根据这个配置文件再生成两个配置文件。

（1）生成 redis-sentinel-26380.conf 配置文件。

```
$ cat sentinel.conf | grep -v "#" | grep -v "^$" > redis-sentinel-26380.conf
$ vi redis-sentinel-26380.conf
```
添加以下内容。
```
port 26380
daemonize yes
pidfile /var/run/redis-sentinel.pid
logfile "26380.log"
dir /tmp
sentinel monitor mymaster 127.0.0.1 7000 2
sentinel down-after-milliseconds mymaster 30000
sentinel parallel-syncs mymaster 1
sentinel failover-timeout mymaster 180000
sentinel deny-scripts-reconfig yes
```
（2）生成 redis-sentinel-26381.conf 配置文件。

```
$ cat sentinel.conf | grep -v "#" | grep -v "^$" > redis-sentinel-26381.conf
$ vi redis-sentinel-26381.conf
```
添加以下内容。
```
port 26381
daemonize yes
pidfile /var/run/redis-sentinel.pid
logfile "26381.log"
```

```
dir /tmp
sentinel monitor mymaster 127.0.0.1 7000 2
sentinel down-after-milliseconds mymaster 30000
sentinel parallel-syncs mymaster 1
sentinel failover-timeout mymaster 180000
sentinel deny-scripts-reconfig yes
```

使用以下命令启动两个 Sentinel 节点。

```
$ redis-sentinel redis-sentinel-26380.conf
$ redis-sentinel redis-sentinel-26381.conf
```

查看 Sentinel 节点的进程。

```
[root@localhost redis-sential]# ps -ef | grep redis-sentinel
root       3215     1  0 00:03 ?        00:00:00 redis-sentinel *:26379 [sentinel]
root       3258     1  0 00:06 ?        00:00:00 redis-sentinel *:26380 [sentinel]
root       3263     1  0 00:06 ?        00:00:00 redis-sentinel *:26381 [sentinel]
root       3276  3050  0 00:07 pts/0    00:00:00 grep --color=auto redis-sentinel
```

使用 redis-cli 命令连接到一个 Sentinel 节点。

```
$ redis-cli -p 26381
```

查看节点信息。

```
[root@localhost redis-sential]# redis-cli -p 26381
127.0.0.1:26381> info sentinel
# Sentinel
sentinel_masters:1
sentinel_tilt:0
sentinel_running_scripts:0
sentinel_scripts_queue_length:0
sentinel_simulate_failure_flags:0
master0:name=mymaster,status=ok,address=127.0.0.1:7000,slaves=2,sentinels=3
```

可以看到 Sentinel 有 3 个，至此 Sentinel 已经配置完成。

使用 redis-cli 命令连接主节点 7000 进行测试。

```
[root@localhost redis-sential]# redis-cli -p 7000
127.0.0.1:7000> SET username xinping
OK
127.0.0.1:7000> SET address beijing
OK
```

再连接从节点 7001。

```
[root@localhost redis-sential]# redis-cli -p 7001
127.0.0.1:7001> KEYS *
1) "username"
2) "address"
127.0.0.1:7001> GET username
"xinping"
127.0.0.1:7001> GET address
"beijing"
```

6.2.3　测试主从切换

1. 查看 Redis 集群信息

在 Redis 客户端使用 info replication 命令查看 Redis 集群信息。

```
[root@localhost redis-sential]# redis-cli -p 7000
127.0.0.1:7000> info replication
# Replication
role:master
connected_slaves:2
slave0:ip=127.0.0.1,port=7001,state=online,offset=503925,lag=0
slave1:ip=127.0.0.1,port=7002,state=online,offset=503925,lag=0
master_replid:5ad27de305049aad423d77c71c79ccf72615af42
master_replid2:0000000000000000000000000000000000000000
master_repl_offset:503925
second_repl_offset:-1
repl_backlog_active:1
repl_backlog_size:1048576
repl_backlog_first_byte_offset:1
repl_backlog_histlen:503925
```

2．查看进程信息

使用 ps -ef | grep redis 命令查看进程信息。

```
[root@localhost redis-sential]# ps -ef | grep redis
root      3191     1  0 00:02 ?        00:00:03 redis-server *:7000
root      3196     1  0 00:02 ?        00:00:03 redis-server *:7001
root      3202     1  0 00:02 ?        00:00:03 redis-server *:7002
root      3215     1  0 00:03 ?        00:00:05 redis-sentinel *:26379 [sentinel]
root      3258     1  0 00:06 ?        00:00:04 redis-sentinel *:26380 [sentinel]
root      3263     1  0 00:06 ?        00:00:04 redis-sentinel *:26381 [sentinel]
root      3696  3050  0 00:48 pts/0    00:00:00 grep --color=auto redis
```

可以看出 Redis 的主节点占用的端口为 3191、3196 和 3202，可以使用 kill 命令结束一个主节点的进程。

使用 kill 命令结果一个主节点的进程。

```
$ kill -9 3191
```

3．查看主节点

使用 redis-cli 命令连接到 7001 节点。

```
[root@localhost data]# redis-cli -p 7001
127.0.0.1:7001> info replication
# Replication
role:master
connected_slaves:1
slave0:ip=127.0.0.1,port=7002,state=online,offset=589803,lag=0
master_replid:dcf84d35e3474035e74f300f351c02aa404efc65
master_replid2:5ad27de305049aad423d77c71c79ccf72615af42
master_repl_offset:589950
second_repl_offset:573709
repl_backlog_active:1
repl_backlog_size:1048576
repl_backlog_first_byte_offset:1
repl_backlog_histlen:589950
```

从上面可以看出主节点变为 7002。

通过哨兵模式的配置，我们可以看出哨兵模式是基于主从模式的，哨兵模式具有主从模式的所有优点。哨兵模式是主从模式的升级，实现了自动化的故障恢复。但哨兵模式的缺点也很明显，Redis 较难实现在线扩容，在集群容量达到上限时在线扩容会变得很复杂。实现哨兵模式的配置也不简单，甚至有些烦琐，于是就有了 Redis 集群。Redis 集群是官方的 Redis 集群实现。

6.3 Redis 集群

Redis 集群是一个由多个主从节点组成的分布式服务器群，它具有复制、高可用和分片特性。Redis 集群将所有数据存储区域划分为 16384 个槽（Slot），每个节点负责一部分槽，槽的信息存储于每个节点中。Redis 集群要将每个节点设置成集群模式，它没有中心节点，可水平扩展，它的性能和高可用性均优于主从模式和哨兵模式，而且集群配置非常简单。Redis 集群架构如图 6-13 所示。

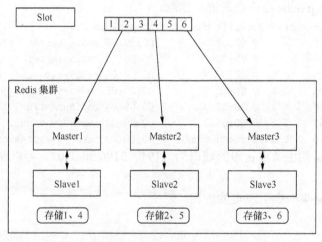

图 6-13 Redis 集群架构

从 Redis 集群架构中可以很容易地看出，首先将数据根据散列规则分配到 6 个槽中，然后根据循环冗余校验（Cyclic Redundancy Check，CRC）算法和取模算法将 6 个槽分别存储到 3 个不同的 Master 节点中，每个 Master 节点又配套部署了一个 Slave 节点，当一个 Master 节点出现问题后，Slave 节点可以顶上。相较于哨兵模式，这种方案的优点在于提高了读写的并发率，分散了 I/O，在保障高可用性的前提下提高了性能。

6.3.1 Redis 集群环境

Redis 集群在物理结构上是由集群上的多个节点构成的，这些节点分为两类，一类叫"主节点"，另一类叫"从节点"。

Redis 集群节点要求如下。

- 主节点不能少于总节点的一半。
- 主节点至少要有 3 个。

一个 Redis 集群正常工作至少需要 3 个主节点且不能少于总节点的一半，本集群环境使用一台节点服务器，在这台服务器开启 6 个 Redis 实例，每个 Redis 实例占用一个端口，模拟 3 个主节点和 3 个从节点环境，组成一个 Redis 集群。本书使用一台服务器部署 6 个 Redis 实例，也可以多台服务器部署 Redis 集群，只修改 Redis 配置文件 redis.conf 的 IP 地址就可以了。本集群实验采用三主三从模式，每个主节点处理各自的数据，提供读写能力，每个从节点异步复制主节点的数据。Redis 5 集群的实验环境如表 6-3 所示。

表 6-3　　　　　　　　　　　Redis 集群的实验环境

IP 地址	端口	版本	操作系统
192.168.11.15	8001	Redis 6.0.6	CentOS 7 64 位
	8002		
	8003		
	8004		
	8005		
	8006		

为了方便配置 Redis 集群，本书以 root 用户登录各 Linux 服务器。

6.3.2　开始 Redis 集群搭建

第 1 步：安装 Redis。

（1）安装 gcc 基础依赖包，使用如下命令。

```
$ yum install gcc-c++
```

（2）下载 Redis 压缩包。

```
$ wget http://download.redis.io/releases/redis-6.0.6.tar.gz
```

解压缩 Redis 压缩包。

```
$ tar -xzvf redis-5.0.5.tar.gz
```

进入解压缩后的目录。

```
$ cd redis-6.0.6/
```

使用 make 命令编译 Redis 源文件。

```
$ make
```

编译成功后，安装 Redis。

```
$ make install PREFIX=/usr/local/redis
```

安装成功后，需要对 Redis 进行部署，把 Redis 的配置文件 redis.conf 复制到/usr/local/redis/conf 目录下。

```
$ mkdir /usr/local/redis/conf
$ cp redis.conf /usr/local/redis/conf
```

（3）配置 Redis 的命令。

最后需要将 Redis 的命令所在目录添加到系统变量 Path 中，修改/etc/profile 文件。

```
$ vi /etc/profile
```

在/etc/profile 文件最后一行添加以下内容。

```
export PATH=$PATH:/usr/local/redis/bin
```

然后使用 source 命令使/etc/profile 文件立即生效。

```
$ source /etc/profile
```

至此，Redis 在 Linux 上的安装和配置就结束了。

第 2 步：在/usr/local 下创建文件夹 redis-cluster，然后在其下创建如下 6 个文件夹。

```
$ mkdir -p /usr/local/redis-cluster
$ cd /usr/local/redis-cluster/
$ mkdir 8001 8002 8003 8004 8005 8006
```

第 3 步：把之前的 redis.conf 配置文件复制到 8001 目录下。

```
$ cp /usr/local/redis/conf/redis.conf /usr/local/redis-cluster/8001
```

使用 vi 命令修改 redis.conf 配置文件。

```
$ vi /usr/local/redis-cluster/8001/redis.conf
```

修改 redis.conf 配置文件中的以下配置项，redis.conf 配置文件可以参考本书的配套实例代码 "Redis\Chapter06\Redis6 集群参考配置文件\redis-cluster\8001\redis.conf"。

（1）daemonize yes：开启 Redis 的守护进程。开启 Redis 的守护进程后，Redis 会在后台一直运行，除非手动输入 kill 命令结束进程。

（2）port 8001：分别对每个节点计算机的端口号进行设置。

（3）dir /usr/local/redis-cluster/8001/：设定数据文件存放位置，必须要指定不同的目录位置，否则会丢失数据。

（4）cluster-enabled yes：启动集群模式。

（5）cluster-config-file nodes-8001.conf：集群节点信息文件，这里 nodes-8001.conf 最好和端口对应。

（6）cluster-node-timeout 5000：集群节点的超时时限，单位为毫秒。

（7）bind 192.168.11.15：修改为主机的 IP 地址，默认 IP 地址为 127.0.0.1，需要修改为其他节点计算机可访问的 IP 地址，否则创建集群时无法访问对应计算机的端口，无法创建集群。

（8）protected-mode no：关闭保护模式。

（9）appendonly yes：开启 AOF 持久化。

如果要设置密码需要增加如下配置。

（10）requirepass xxx：设置 Redis 的访问密码（本例使用 xxx 代替密码，读者可根据实际情况自行设置）。

（11）masterauth xxx：设置集群节点间的访问密码（本例使用 xxx 代替密码，读者可根据实际情况自行设置），与（10）中设置的密码一致。

第 4 步：把修改后的 redis.conf 配置文件复制到 8002、8003、8004、8005 和 8006 目录下，修改（2）（3）（5）里的端口。

第 3 步已经完成了一个 Redis 节点的配置，接下来就是机械化地再完成另外 5 个节点的配置。其实可以这么做：把 8001 实例的 redis.conf 配置文件复制到另外 5 个文件夹中，并修改 redis.conf 配置文件中所有和端口相关的信息，其实就是 port，dir 和 cluster-config-file 配置项的端口信息。

```
$ cd /usr/local/redis-cluster/8001
$ cp redis.conf /usr/local/redis-cluster/8002
$ cp redis.conf /usr/local/redis-cluster/8003
$ cp redis.conf /usr/local/redis-cluster/8004
$ cp redis.conf /usr/local/redis-cluster/8005
$ cp redis.conf /usr/local/redis-cluster/8006
```

第 5 步：分别启动 6 个 Redis 节点，然后检查 Redis 节点是否启动成功。

```
$ redis-server /usr/local/redis-cluster/8001/redis.conf
$ redis-server /usr/local/redis-cluster/8002/redis.conf
$ redis-server /usr/local/redis-cluster/8003/redis.conf
$ redis-server /usr/local/redis-cluster/8004/redis.conf
$ redis-server /usr/local/redis-cluster/8005/redis.conf
$ redis-server /usr/local/redis-cluster/8006/redis.conf
```

执行启动 6 个 Redis 节点命令的返回结果如图 6-14 所示。

```
[root@localhost 8006]# redis-server /usr/local/redis-cluster/8001/redis.conf
10468:C 02 Oct 2019 00:16:21.352 # oO0OoO0OoO0Oo Redis is starting oO0OoO0OoO0Oo
10468:C 02 Oct 2019 00:16:21.352 # Redis version=5.0.5, bits=64, commit=00000000, modified=0, pid=10468, just started
10468:C 02 Oct 2019 00:16:21.352 # Configuration loaded
[root@localhost 8006]# redis-server /usr/local/redis-cluster/8002/redis.conf
10473:C 02 Oct 2019 00:16:31.190 # oO0OoO0OoO0Oo Redis is starting oO0OoO0OoO0Oo
10473:C 02 Oct 2019 00:16:31.190 # Redis version=5.0.5, bits=64, commit=00000000, modified=0, pid=10473, just started
10473:C 02 Oct 2019 00:16:31.190 # Configuration loaded
[root@localhost 8006]# redis-server /usr/local/redis-cluster/8003/redis.conf
10478:C 02 Oct 2019 00:16:38.516 # oO0OoO0OoO0Oo Redis is starting oO0OoO0OoO0Oo
10478:C 02 Oct 2019 00:16:38.516 # Redis version=5.0.5, bits=64, commit=00000000, modified=0, pid=10478, just started
10478:C 02 Oct 2019 00:16:38.516 # Configuration loaded
[root@localhost 8006]# redis-server /usr/local/redis-cluster/8004/redis.conf
10483:C 02 Oct 2019 00:16:44.502 # oO0OoO0OoO0Oo Redis is starting oO0OoO0OoO0Oo
10483:C 02 Oct 2019 00:16:44.502 # Redis version=5.0.5, bits=64, commit=00000000, modified=0, pid=10483, just started
10483:C 02 Oct 2019 00:16:44.502 # Configuration loaded
[root@localhost 8006]# redis-server /usr/local/redis-cluster/8005/redis.conf
10488:C 02 Oct 2019 00:16:48.679 # oO0OoO0OoO0Oo Redis is starting oO0OoO0OoO0Oo
10488:C 02 Oct 2019 00:16:48.679 # Redis version=5.0.5, bits=64, commit=00000000, modified=0, pid=10488, just started
10488:C 02 Oct 2019 00:16:48.679 # Configuration loaded
[root@localhost 8006]# redis-server /usr/local/redis-cluster/8006/redis.conf
10501:C 02 Oct 2019 00:16:52.869 # oO0OoO0OoO0Oo Redis is starting oO0OoO0OoO0Oo
10501:C 02 Oct 2019 00:16:52.869 # Redis version=5.0.5, bits=64, commit=00000000, modified=0, pid=10501, just started
10501:C 02 Oct 2019 00:16:52.869 # Configuration loaded
```

图 6-14　启动 6 个 Redis 节点

使用 ps -ef | grep redis 命令查看 Redis 节点是否启动成功。从图 6-15 可以看出 6 个 Redis 节点已经全部启动成功了。

```
[root@localhost ~]# ps -ef | grep redis
root      2573     1  0 06:55 ?        00:00:00 redis-server 192.168.11.15:8001 [cluster]
root      2579     1  0 06:55 ?        00:00:00 redis-server 192.168.11.15:8002 [cluster]
root      2585     1  0 06:55 ?        00:00:00 redis-server 192.168.11.15:8003 [cluster]
root      2591     1  0 06:55 ?        00:00:00 redis-server 192.168.11.15:8004 [cluster]
root      2597     1  0 06:55 ?        00:00:00 redis-server 192.168.11.15:8005 [cluster]
root      2603     1  0 06:55 ?        00:00:00 redis-server 192.168.11.15:8006 [cluster]
root      2680  2474  0 06:58 pts/0    00:00:00 grep --color=auto redis
```

图 6-15　查看 Redis 节点的进程

使用命令一个个启动 Redis 节点比较麻烦，我们可以使用脚本的方式启动 Redis 节点，使用以下命令创建 Redis 集群启动脚本（startRedisCluster.sh）。

```
$ touch startRedisCluster.sh
$ chmod +x startRedisCluster.sh
```

使用 vi startRedisCluster.sh 命令修改 startRedisCluster.sh 脚本，添加以下内容。

```
redis-server /usr/local/redis-cluster/8001/redis.conf
redis-server /usr/local/redis-cluster/8002/redis.conf
redis-server /usr/local/redis-cluster/8003/redis.conf
redis-server /usr/local/redis-cluster/8004/redis.conf
redis-server /usr/local/redis-cluster/8005/redis.conf
redis-server /usr/local/redis-cluster/8006/redis.conf
```

使用以下命令启动脚本。

```
$ ./startRedisCluster.sh
```

第 6 步：使用 redis-cli 命令创建 Redis 集群，如图 6-16 所示。

```
$ redis-cli --cluster create --cluster-replicas 1 192.168.11.15:8001 192.168.11.
15:8002 192.168.11.15:8003 192.168.11.15:8004 192.168.11.15:8005 192.168.11.15:8006
```

使用 redis-cli 命令创建 Redis 集群使用的参数--cluster-replicas 表示主节点和从节点的比例，当参数--cluster-replicas 为 1 时，表示创建 Redis 集群时一个主节点需要有一个从节点。

按照提示“Can I set the above configuration?”输入 yes。

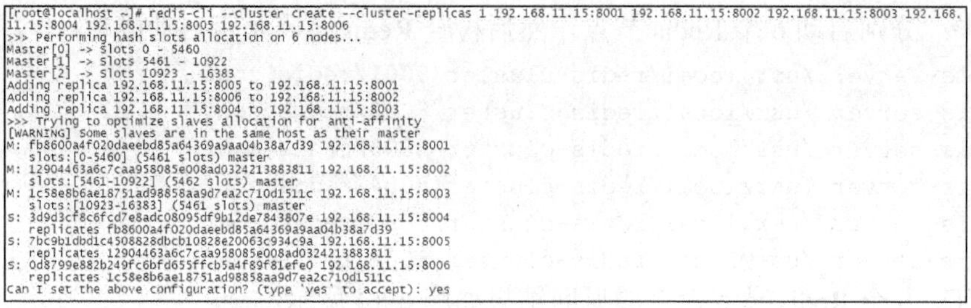

```
[root@localhost ~]# redis-cli --cluster create --cluster-replicas 1 192.168.11.15:8001 192.168.11.15:8002 192.168.11.15:8003 192.168.
11.15:8004 192.168.11.15:8005 192.168.11.15:8006
>>> Performing hash slots allocation on 6 nodes...
Master[0] -> Slots 0 - 5460
Master[1] -> Slots 5461 - 10922
Master[2] -> Slots 10923 - 16383
Adding replica 192.168.11.15:8005 to 192.168.11.15:8001
Adding replica 192.168.11.15:8006 to 192.168.11.15:8002
Adding replica 192.168.11.15:8004 to 192.168.11.15:8003
>>> Trying to optimize slaves allocation for anti-affinity
[WARNING] Some slaves are in the same host as their master
M: fb8600a4f020daeebd85a64369a9aa04b38a7d39 192.168.11.15:8001
   slots:[0-5460] (5461 slots) master
M: 12904463a6c7caa958085e008ad0324213883811 192.168.11.15:8002
   slots:[5461-10922] (5462 slots) master
M: 1c58e8b6ae18751ad98858aa9d7ea2c710d1511c 192.168.11.15:8003
   slots:[10923-16383] (5461 slots) master
S: 3d9d3cf8c6fcd7e8adc08095df9b12de7843807e 192.168.11.15:8004
   replicates fb8600a4f020daeebd85a64369a9aa04b38a7d39
S: 7bc9b1dbd1c4508828dbcb10828e20063c934c9a 192.168.11.15:8005
   replicates 12904463a6c7caa958085e008ad0324213883811
S: 0d8799e882b249fc6bfd655ffcb5a4f89f81efe0 192.168.11.15:8006
   replicates 1c58e8b6ae18751ad98858aa9d7ea2c710d1511c
Can I set the above configuration? (type 'yes' to accept): yes
```

图 6-16 创建 Redis 集群

```
Can I set the above configuration? (type 'yes' to accept): yes
>>> Nodes configuration updated
>>> Assign a different config epoch to each node
>>> Sending CLUSTER MEET messages to join the cluster
Waiting for the cluster to join

>>> Performing Cluster Check (using node 192.168.11.15:8001)
M: fb8600a4f020daeebd85a64369a9aa04b38a7d39 192.168.11.15:8001
   slots:[0-5460] (5461 slots) master
   1 additional replica(s)
S: 3d9d3cf8c6fcd7e8adc08095df9b12de7843807e 192.168.11.15:8004
   slots: (0 slots) slave
   replicates fb8600a4f020daeebd85a64369a9aa04b38a7d39
M: 12904463a6c7caa958085e008ad0324213883811 192.168.11.15:8002
   slots:[5461-10922] (5462 slots) master
   1 additional replica(s)
M: 1c58e8b6ae18751ad98858aa9d7ea2c710d1511c 192.168.11.15:8003
   slots:[10923-16383] (5461 slots) master
   1 additional replica(s)
S: 0d8799e882b249fc6bfd655ffcb5a4f89f81efe0 192.168.11.15:8006
   slots: (0 slots) slave
   replicates 1c58e8b6ae18751ad98858aa9d7ea2c710d1511c
S: 7bc9b1dbd1c4508828dbcb10828e20063c934c9a 192.168.11.15:8005
   slots: (0 slots) slave
   replicates 12904463a6c7caa958085e008ad0324213883811
[OK] All nodes agree about slots configuration.
>>> Check for open slots...
>>> Check slots coverage...
[OK] All 16384 slots covered.
```

以上是创建的 Redis 集群信息，可以看出每创建一个 Redis 主节点就创建一个 Redis 从节点。上述信息可以转换为表格形式，如表 6-4 所示。

表 6-4　　　　　　　　Redis 集群的端口、角色、节点 ID 和槽范围

端口	角色	节点 ID	槽范围
8001	Master	fb8600a4f020daeebd85a64369a9aa04b38a7d39	0～5460
8004	Slave	3d9d3cf8c6fcd7e8adc08095df9b12de7843807e	0
8002	Master	12904463a6c7caa958085e008ad0324213883811	5461～10922
8005	Slave	7bc9b1dbd1c4508828dbcb10828e20063c934c9a	0
8003	Master	1c58e8b6ae18751ad98858aa9d7ea2c710d1511c	10923～16383
8006	Slave	0d8799e882b249fc6bfd655ffcb5a4f89f81efe0	0

Redis 集群的主从节点关系如图 6-13 所示，图中 Redis 集群有三个主节点和三个从节点，一个主节点对应一个从节点，形成一对一的对应关系，如图 6-17 所示。

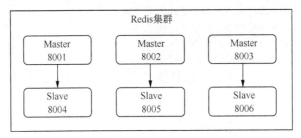

图 6-17　Redis 集群的主从节点关系

第 7 步：验证 Redis 集群。

（1）使用 redis-cli 命令连接 Redis 集群节点，实例如下。

`$ redis-cli -c -a　　-h 192.168.11.15 -p 8001`

redis-cli 命令的参数-a 表示访问 Redis 服务器密码，-c 表示连接 Redis 集群，-h 表示 Redis 集群节点的 IP 地址，-p 表示 Redis 集群节点的端口号。如果 Redis 集群没有设置密码，可以省略参数-a。

可以使用以下任意一条命令连接 Redis 集群中的某个节点。

```
$ redis-cli -c -h 192.168.11.15 -p 8001
$ redis-cli -c -h 192.168.11.15 -p 8002
$ redis-cli -c -h 192.168.11.15 -p 8003
$ redis-cli -c -h 192.168.11.15 -p 8004
$ redis-cli -c -h 192.168.11.15 -p 8005
$ redis-cli -c -h 192.168.11.15 -p 8006
```

例如，使用命令 redis-cli -c -h 192.168.11.15 -p 8001 访问 Redis 集群。

```
[root@localhost 8006]# redis-cli -c -h 192.168.11.15 -p 8001
192.168.11.15:8001> SET name xinping
-> Redirected to slot [5798] located at 192.168.11.15:8002
OK
192.168.11.15:8002> get name
"xinping"
```

使用 SET 命令后，进入了端口为 8002 的 Redis 节点，进行了跳转重定向（Redirected）。

（2）测试 Redis 集群是否正常。

使用 cluster info 命令查看 Redis 集群信息，返回值"cluster_known_nodes:6"表示当前 Redis 集群中共有 6 个 Redis 节点。

```
192.168.11.15:8001> cluster info
cluster_state:ok
cluster_slots_assigned:16384
cluster_slots_ok:16384
cluster_slots_pfail:0
cluster_slots_fail:0
cluster_known_nodes:6
cluster_size:3
cluster_current_epoch:6
cluster_my_epoch:1
```

```
cluster_stats_messages_ping_sent:9036
cluster_stats_messages_pong_sent:9029
cluster_stats_messages_sent:18065
cluster_stats_messages_ping_received:9024
cluster_stats_messages_pong_received:9036
cluster_stats_messages_meet_received:5
cluster_stats_messages_received:18065
```

使用 cluster nodes 命令查看 Redis 集群节点列表，可以看出整个 Redis 集群含有 3 个主节点和 3 个从节点。

```
192.168.11.15:8001> cluster nodes
    3d9d3cf8c6fcd7e8adc08095df9b12de7843807e 192.168.11.15:8004@18004 slave
fb8600a4f020daeebd85a64369a9aa04b38a7d39 0 1598792307000 1 connected
    12904463a6c7caa958085e008ad0324213883811 192.168.11.15:8002@18002 master -
0 1598792309022 2 connected 5461-10922
    fb8600a4f020daeebd85a64369a9aa04b38a7d39 192.168.11.15:8001@18001 myself,master -
0 1598792307000 1 connected 0-5460
    1c58e8b6ae18751ad98858aa9d7ea2c710d1511c 192.168.11.15:8003@18003 master -
0 1598792307274 3 connected 10923-16383
    0d8799e882b249fc6bfd655ffcb5a4f89f81efe0 192.168.11.15:8006@18006 slave
1c58e8b6ae18751ad98858aa9d7ea2c710d1511c 0 1598792307478 3 connected
    7bc9b1dbd1c4508828dbcb10828e20063c934c9a 192.168.11.15:8005@18005 slave
12904463a6c7caa958085e008ad0324213883811 0 1598792308506 2 connected
```

至此，Redis 集群环境搭建完毕。

（3）使用 redis-cli 命令关闭 Redis 集群节点，实例如下。

```
$ redis-cli -a xxx -c -h 192.168.11.15 -p 8001 shutdown
```

在本例中可以使用以下命令关闭 6 个 Redis 集群节点。

```
$ redis-cli -c -h 192.168.11.15 -p 8001 shutdown
$ redis-cli -c -h 192.168.11.15 -p 8002 shutdown
$ redis-cli -c -h 192.168.11.15 -p 8003 shutdown
$ redis-cli -c -h 192.168.11.15 -p 8004 shutdown
$ redis-cli -c -h 192.168.11.15 -p 8005 shutdown
$ redis-cli -c -h 192.168.11.15 -p 8006 shutdown
```

在此做一个实验，使用以下命令关闭端口为 8001 的 Redis 集群节点。

```
$ redis-cli -c -h 192.168.11.15 -p 8001 shutdown
```

再查看 Redis 进程，会发现端口为 8001 的 Redis 进程已经销毁。

```
[root@localhost ~]# ps -ef | grep redis
root      16478     1  0 20:44 ?        00:00:01 redis-server 192.168.11.15:8001 [cluster]
root      16484     1  0 20:44 ?        00:00:01 redis-server 192.168.11.15:8002 [cluster]
root      16490     1  0 20:44 ?        00:00:01 redis-server 192.168.11.15:8003 [cluster]
root      16496     1  0 20:44 ?        00:00:01 redis-server 192.168.11.15:8004 [cluster]
root      16502     1  0 20:44 ?        00:00:01 redis-server 192.168.11.15:8005 [cluster]
root      16508     1  0 20:44 ?        00:00:01 redis-server 192.168.11.15:8006 [cluster]
root      16688 16645  0 20:58 pts/5    00:00:00 redis-cli -c -h 192.168.11.15 -p 8001
root      16740 16400  0 21:01 pts/4    00:00:00 grep --color=auto redis
```

Redis 客户端在连接上端口为 8002 的 Redis 集群节点后，查看 Redis 集群节点信息，会发现端口为 8001 的 Redis 集群节点的状态为 "disconnected"，表示这个集群节点已经关闭。

```
[root@localhost ~]# redis-cli -c -h 192.168.11.15 -p 8002
192.168.11.15:8002> cluster nodes
```

```
    0d8799e882b249fc6bfd655ffcb5a4f89f81efe0 192.168.11.15:8006@18006 slave
1c58e8b6ae18751ad98858aa9d7ea2c710d1511c 0 1598792578903 3 connected
    12904463a6c7caa958085e008ad0324213883811 192.168.11.15:8002@18002 myself,master -
0 1598792576000 2 connected 5461-10922
    3d9d3cf8c6fcd7e8adc08095df9b12de7843807e 192.168.11.15:8004@18004 master -
0 1598792576550 7 connected 0-5460
    fb8600a4f020daeebd85a64369a9aa04b38a7d39 192.168.11.15:8001@18001 master,fail -
1598792556868 1598792554304 1 disconnected
    1c58e8b6ae18751ad98858aa9d7ea2c710d1511c 192.168.11.15:8003@18003 master -
0 1598792578000 3 connected 10923-16383
    7bc9b1dbd1c4508828dbcb10828e20063c934c9a 192.168.11.15:8005@18005 slave
12904463a6c7caa958085e008ad0324213883811 0 1598792576860 2 connected
```

如果想要打开端口为 8001 的 Redis 集群节点，需要使用如下命令。

```
$ redis-server /usr/local/redis-cluster/8001/redis.conf
```

6.3.3 Redis 集群代理

集群代理（Cluster Proxy）是 Redis 6 的新特性。Redis 集群代理（Redis Cluster Proxy）允许 Redis 客户端不需要知道集群中的具体节点个数和主从身份，直接通过集群代理访问集群。对于客户端来说，通过集群代理访问集群就和访问单机的 Redis 服务器一样，可以解除很多集群的使用限制。Redis 集群代理架构如图 6-18 所示。

Redis 集群代理使用的服务器资源如表 6-5 所示。使用上一节已经配置好的 Redis 集群环境，将 Redis 集群代理安装在 CentOS 7 操作系统下。

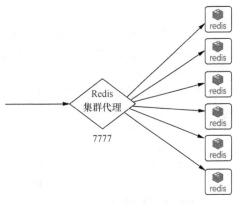

图 6-18 Redis 集群代理架构

Redis 集群代理目前的最新版本（截至 2020 年 8 月 17 日）是 1.0-beta2，是测试版本，包含相对较多的漏洞，请谨慎在生产环境下使用，期待后续有稳定版本推出。

表 6-5　　　　　　　　　　　Redis 集群代理实验环境

IP 地址	端口	版本	操作系统
192.168.11.15	8001	Redis 6.0.6	CentOS 7 64 位
	8002		
	8003		
	8004		
	8005		
	8006		
	7777	Redis Cluster Proxy 1.0-beta2	

以 root 用户登录 Linux 服务器。

使用 Redis 集群代理的实例如下。

1. 安装 Redis 集群代理

（1）安装 gcc 9.3.1。

首先，使用如下命令安装 gcc 基础依赖包。

```
$ yum -y install gcc-c++
```

为了编译 Redis Cluster Proxy1.0-beta2 源码还需要使用 devtoolset 命令升级 gcc。gcc 版本必须在 4.9 以上。

```
$ yum -y install centos-release-scl
$ yum -y install devtoolset-9-gcc devtoolset-9-gcc-c++ devtoolset-9-binutils
$ scl enable devtoolset-9 bash
$ echo "source /opt/rh/devtoolset-9/enable" >>/etc/profile
```

安装完 devtoolset 后，需要输入 scl enable devtoolset-9 bash 命令来启动 devtoolset。启动 devtoolset 的命令仅针对本次会话有效，若重新登录 Linux，需要再次使用 scl 命令启动 devtoolset。若要使 devtoolset 长期有效，需要输入 echo "source /opt/rh/devtoolset-9/enable" >>/etc/profile 命令。

（2）下载、编译、安装和配置 Redis 集群代理。

然后安装 git。git 是一个分布式版本控制系统。

```
$ yum install git
```

建立下载目录/upload，并在这个目录下使用 git 下载最新版本的 Redis Cluster Proxy 源码包。

```
$ mkdir /upload
$ cd /upload
$ git clone https://github.com/artix75/redis-cluster-proxy
```

下载完成后，进入创建的 redis-cluster-proxy 目录。

```
$ cd redis-cluster-proxy
```

使用 make 命令安装 Redis Cluster Proxy。

```
$ make PREFIX=/usr/local/redis_cluster_proxy install
```

make 命令的参数 PREFIX 要大写，代表安装路径。执行 make 命令后，Redis Cluster Proxy 的可执行命令会被自动复制到/usr/local/redis_cluster_proxy /bin 目录下，这样执行 Redis Cluster Proxy 命令时，就不用输入完整路径了。

安装成功后，需要对 Redis Cluster Proxy 进行部署。把它的配置文件 proxy.conf 复制到 /usr/local/redis-cluster-proxy/conf 目录下。

```
$ mkdir -p /usr/local/redis_cluster_proxy/conf
$ cd /upload/redis-cluster-proxy
$ cp proxy.conf /usr/local/redis_cluster_proxy/conf
```

部署后，Redis 集群代理的目录结构如图 6-19 所示。

使用 vi 命令修改 Redis 集群代理的配置文件 proxy.conf。

```
vi /usr/local/redis_cluster_proxy/conf/proxy.conf
```

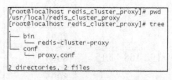

图 6-19 集群代理的目录结构

修改 proxy.conf 配置文件中的以下配置项。

```
#    Redis                           7777
port 7777

#    Redis
```

```
daemonize yes

#     Redis
logfile "/usr/local/redis_cluster_proxy/conf/redis-cluster-proxy.log"

#     Redis
cluster 192.168.11.15:8001
cluster 192.168.11.15:8002
cluster 192.168.11.15:8003
cluster 192.168.11.15:8004
cluster 192.168.11.15:8005
cluster 192.168.11.15:8006
```

最后，将 Redis Cluster Proxy 的可执行命令所在目录添加到系统参数 PATH 中，修改/etc/profile 文件。

```
$ vi /etc/profile
```

在/etc/profile 文件最后一行添加如下内容。

```
export PATH=$PATH:/usr/local/redis_cluster_proxy/bin
```

然后使用 source 命令使这个文件立即生效。

```
$ source /etc/profile
```

至此，Redis 集群代理在 Linux 上的安装和配置就结束了。

2. 启动和使用 Redis 集群代理

安装并配置好 Redis Cluster Proxy 后，启动 Redis 集群代理。

```
$ redis-cluster-proxy -c /usr/local/redis_cluster_proxy/conf/proxy.conf
```

redis-cluster-proxy 命令的参数-c 表示加载 proxy.conf 配置文件后启动 Redis 集群代理。执行启动 Redis 集群代理命令的返回结果如图 6-20 所示，可以看出 Redis 集群代理已经成功启动。

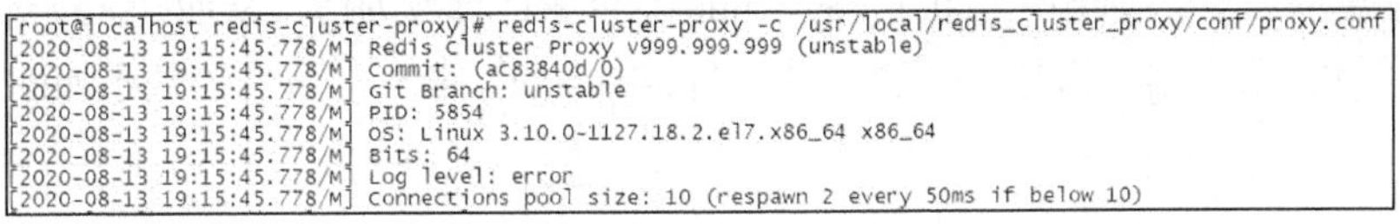

```
[root@localhost redis-cluster-proxy]# redis-cluster-proxy -c /usr/local/redis_cluster_proxy/conf/proxy.conf
[2020-08-13 19:15:45.778/M] Redis Cluster Proxy v999.999.999 (unstable)
[2020-08-13 19:15:45.778/M] Commit: (ac83840d/0)
[2020-08-13 19:15:45.778/M] Git Branch: unstable
[2020-08-13 19:15:45.778/M] PID: 5854
[2020-08-13 19:15:45.778/M] OS: Linux 3.10.0-1127.18.2.el7.x86_64 x86_64
[2020-08-13 19:15:45.778/M] Bits: 64
[2020-08-13 19:15:45.778/M] Log level: error
[2020-08-13 19:15:45.778/M] Connections pool size: 10 (respawn 2 every 50ms if below 10)
```

图 6-20　启动 Redis 集群代理

执行 ps -ef | grep redis 命令，返回结果如图 6-21 所示，可以看出 Redis 集群代理进程占用的端口是 5855。

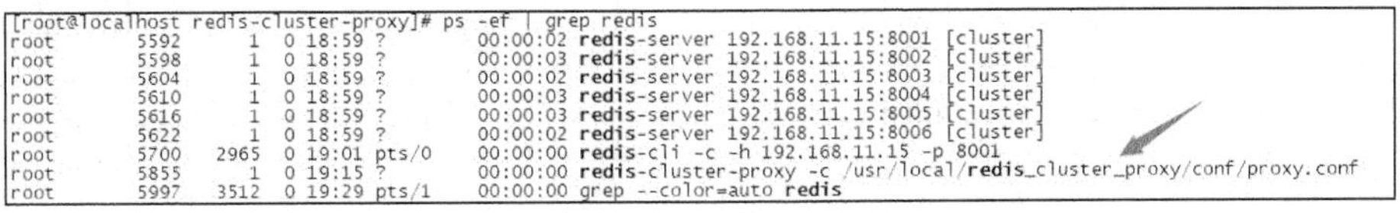

```
[root@localhost redis-cluster-proxy]# ps -ef | grep redis
root      5592     1  0 18:59 ?        00:00:02 redis-server 192.168.11.15:8001 [cluster]
root      5598     1  0 18:59 ?        00:00:03 redis-server 192.168.11.15:8002 [cluster]
root      5604     1  0 18:59 ?        00:00:02 redis-server 192.168.11.15:8003 [cluster]
root      5610     1  0 18:59 ?        00:00:03 redis-server 192.168.11.15:8004 [cluster]
root      5616     1  0 18:59 ?        00:00:03 redis-server 192.168.11.15:8005 [cluster]
root      5622     1  0 18:59 ?        00:00:02 redis-server 192.168.11.15:8006 [cluster]
root      5700  2965  0 19:01 pts/0    00:00:00 redis-cli -c -h 192.168.11.15 -p 8001
root      5855     1  0 19:15 ?        00:00:00 redis-cluster-proxy -c /usr/local/redis_cluster_proxy/conf/proxy.conf
root      5997  3512  0 19:29 pts/1    00:00:00 grep --color=auto redis
```

图 6-21　查看 Redis 集群代理进程占用的端口

如果需要停止 Redis 集群代理，可以使用 kill 命令无条件终止 Redis 集群代理进程。

```
$ kill -9 5855
```

在 Redis 客户端使用如下命令连接到 Redis 集群代理。

```
$ redis-cli -h 192.168.11.15 -p 7777
```

与 Redis 集群连接方式不同，在 Redis 集群代理模式下，Redis 客户端可以连接至 Redis

集群代理节点，而无须知道 Redis 集群自身的详细信息，操作 Redis 集群和操作单机 Redis 服务器是一样的，通过 Redis 集群代理访问并操作 Redis 集群的实例如图 6-22 所示。

使用传统的 Redis 集群连接方式来查看上面 SADD 操作的结果如图 6-23 所示。可以发现数据的确是写入 Redis 集群中不同的节点中了。

图 6-22 通过 Redis 集群代理访问 Redis 集群　　图 6-23 通过 Redis 客户端访问 Redis 集群

6.3.4　Redis 集群特点

Redis 集群方案采用的是散列分区的"虚拟槽分区"方式，槽范围是 0～16383，共用 16384（即 2^{14}）个槽。每个节点会维护自身负责的槽及槽所映射的键值对数据。

所有 key 数据散列函数 CRC16（key）%16384（可使用按位与操作优化为 CRC16（key）&16383）被映射到槽内。具体来说，当需要在 Redis 集群中存储一个 key-value 对时，Redis 先对 key 使用 CRC16 算法算出一个循环冗余校验码，然后将校验码对 16384 取余。这样每个 key 都会有一个 0～16383 的余数，从而对应一个槽。Redis 会根据数量大致均等的原则将槽映射到不同的节点。例如，有 3 个主节点时，每个节点大致负责 5500 个槽的读写。Redis 集群使用自己设计的简单散列算法 CRC16（key）%16384，而非一致性散列算法。Redis 的作者认为简单散列算法虽然没有一致性散列算法灵活，但效果已经不错了，且实现很简单，增删节点处理起来也很方便。为了与一致性散列算法区别开来，使用简单散列算法的 Redis 槽一般称为散列槽。

在 Redis 集群的实际使用中，SET key value 命令会计算散列值，从而把 key-value 对存储到对应的散列槽和主节点中。槽与节点的关系如图 6-24 所示，key 与数据的关系如图 6-25 所示。

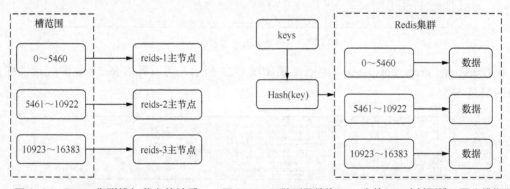

图 6-24　Redis 集群槽与节点的关系　　图 6-25　用散列函数将 keys 中的 key 映射到槽，再由槽指向数据

启动 Redis 集群后，使用 Redis 客户端连接服务器 192.168.11.15 上端口为 8001 的 Redis 集群节点。

```
$ redis-cli -c -h 192.168.11.15 -p 8001
192.168.11.15:8001>
```

然后执行 cluster nodes 命令查看当前 Redis 集群的节点信息。

```
[root@localhost ~]# redis-cli -c -h 192.168.11.15 -p 8001
192.168.11.15:8001> cluster nodes
    7bc9b1dbd1c4508828dbcb10828e20063c934c9a 192.168.11.15:8005@18005 slave
12904463a6c7caa958085e008ad0324213883811 0 1598793006608 2 connected
    0d8799e882b249fc6bfd655ffcb5a4f89f81efe0 192.168.11.15:8006@18006 slave
1c58e8b6ae18751ad98858aa9d7ea2c710d1511c 0 1598793006000 3 connected
    12904463a6c7caa958085e008ad0324213883811 192.168.11.15:8002@18002 master -
0 1598793005587 2 connected 5461-10922
    1c58e8b6ae18751ad98858aa9d7ea2c710d1511c 192.168.11.15:8003@18003 master -
0 1598793006000 3 connected 10923-16383
    fb8600a4f020daeebd85a64369a9aa04b38a7d39 192.168.11.15:8001@18001 myself,slave
3d9d3cf8c6fcd7e8adc08095df9b12de7843807e 0 1598793005000 7 connected
    3d9d3cf8c6fcd7e8adc08095df9b12de7843807e 192.168.11.15:8004@18004 master -
0 1598793006000 7 connected 0-5460
```

该命令返回值显示了 Redis 集群中的每个节点的 ID、身份、连接数和槽数等信息。从返回值可以看出，整个 Redis 集群运行正常，其中包含三个主节点和三个从节点。

- 8001 端口的 Redis 集群主节点存储 0～5460 的散列槽。
- 8002 端口的 Redis 集群主节点存储 5461～10922 的散列槽。
- 8003 端口的 Redis 集群主节点存储 10923～16383 的散列槽。

这三个主节点存储的所有槽组成 Redis 集群的存储槽位。从节点是主节点的备份，不显示存储槽位。

使用散列槽可以方便地增加或移除节点。当需要增加节点时，只需要把其他节点的某些散列槽挪到新节点就可以了。当需要移除节点时，只需要把待移除节点上的散列槽挪到其他节点就可以了。增加或移除节点的时候不需要先停止所有 Redis 服务。

一个 Redis 集群总共有 16384 个散列槽，一个散列槽中会有很多 key-value 对。这一结构可以理解成表的分区：使用单机的 redis 时只有一个表，所有的 key 都放在这个表里；改用 Redis 集群以后会自动生成 16384 个分区表，增加数据时根据简单散列算法来决定 key 应该存储在哪个分区，每个分区可以存储很多 key。

可以使用 cluster info 命令查看 Redis 集群信息。

```
192.168.11.15:8005> cluster info
cluster_state:ok
cluster_slots_assigned:16384
cluster_slots_ok:16384
cluster_slots_pfail:0
cluster_slots_fail:0
cluster_known_nodes:6
cluster_size:3
cluster_current_epoch:7
cluster_my_epoch:7
cluster_stats_messages_ping_sent:1225
cluster_stats_messages_pong_sent:1253
cluster_stats_messages_sent:2478
cluster_stats_messages_ping_received:1253
cluster_stats_messages_pong_received:1224
cluster_stats_messages_fail_received:1
cluster_stats_messages_received:2478
```

当我们执行 SET age 23 命令时，Redis 是如何将数据保存到集群中的呢？

```
192.168.11.15:8001> SET age 23
-> Redirected to slot [741] located at 192.168.11.15:8005
OK
192.168.11.15:8005>
```

输入 SET age 23 命令后，Redis 的执行步骤如下。

（1）接收命令 SET age 23。

（2）通过 key（age）计算出对应的槽，然后根据槽找到对应的节点。（age 对应的槽为 741）

（3）重定向到对应的节点执行命令。

整个 Redis 集群提供了 16384 个槽，也就是说集群中各节点分得的槽数总和为 16384。Redis 集群实现了将 16384 个槽平均分配给了 N 个节点，如图 6-26 所示。

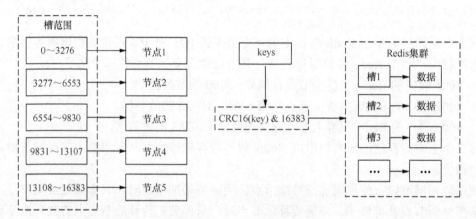

图 6-26　N=5 时虚拟槽分配

 如果有部分槽没有指定到 Redis 节点，那么这部分槽所对应的 keys 将不能使用。Redis 集群的数据分散度高，键值分布与业务无关，键值无法顺序访问，支持批量操作。

6.3.5　新增 Redis 集群节点

在 6.3.2 小节中我们把 6 个 Redis 节点部署在一台 Linux 服务器上，采用三主三从的模式启动 Redis 集群。本小节我们进行 Redis 集群的水平扩展实验，在原始 Redis 集群基础上新增两个 Redis 节点，一个主节点（端口为 8007）和一个从节点（端口为 8008）。Redis 集群新增节点的架构如图 6-27 所示。新增 Redis 节点的顺序是先增加主节点，然后再增加从节点。

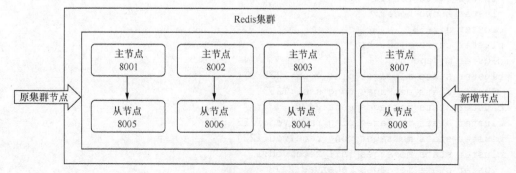

图 6-27　Redis 集群新增节点

1. 新增 Redis 主节点

（1）新增 Redis 节点

新增一个 Redis 节点，需要在/usr/local/redis-cluster 目录下创建相应的文件夹。本例执行以下操作创建 8007 文件夹。

```
$ cd /usr/local/redis-cluster/
$ mkdir 8007
```

复制 8001 文件夹下的 redis.conf 配置文件到 8007 文件夹下。

```
cp /usr/local/redis/conf/redis.conf /usr/local/redis-cluster/8007
```

修改 redis.conf 配置文件的配置项。

```
#     Redis
daemonize  yes

#     Redis
port   8007

#
dir   /usr/local/redis-cluster/8007/

#
cluster-enabled  yes

#
cluster-config-file  nodes-8007.conf

#
cluster-node-timeout  5000

#      Linux       IP
bind  192.168.11.15

#
protected-mode  no

#     AOF        AOF         Redis
appendonly  yes
```

（2）启动新增的 Redis 节点

启动端口为 8007 的 Redis 节点。

```
[root@localhost ~]# redis-server /usr/local/redis-cluster/8007/redis.conf
16904:C 30 Aug 2020 21:13:34.953 # oO0OoO0OoO0Oo Redis is starting oO0OoO0OoO0Oo
16904:C 30 Aug 2020 21:13:34.953 # Redis version=6.0.6, bits=64, commit=00000000, modified=0, pid=16904, just started
16904:C 30 Aug 2020 21:13:34.953 # Configuration loaded
```

查看 Redis 节点的进程启动情况，从图 6-28 可以看出端口为 8007 的 Redis 节点已经启动成功。

这时查看 Redis 集群的节点信息，会发现并没有端口为 8007 的 Redis 节点信息。还需要增加 Redis 主节点。

```
[root@localhost ~]# ps -ef | grep redis
root      16484      1  0 20:44 ?        00:00:03 redis-server 192.168.11.15:8002 [cluster]
root      16490      1  0 20:44 ?        00:00:03 redis-server 192.168.11.15:8003 [cluster]
root      16496      1  0 20:44 ?        00:00:03 redis-server 192.168.11.15:8004 [cluster]
root      16502      1  0 20:44 ?        00:00:03 redis-server 192.168.11.15:8005 [cluster]
root      16508      1  0 20:44 ?        00:00:03 redis-server 192.168.11.15:8006 [cluster]
root      16805      1  0 21:03 ?        00:00:01 redis-server 192.168.11.15:8001 [cluster]
root      16869  16645  0 21:10 pts/5    00:00:00 redis-cli -c -h 192.168.11.15 -p 8001
root      16905      1  0 21:13 ?        00:00:00 redis-server 192.168.11.15:8007 [cluster]
root      16912  16400  0 21:14 pts/4    00:00:00 grep --color=auto redis
```

图 6-28 查看 Redis 节点进程

```
[root@localhost ~]# redis-cli -c -h 192.168.11.15 -p 8001
192.168.11.15:8001> cluster nodes
7bc9b1dbd1c4508828dbcb10828e20063c934c9a 192.168.11.15:8005@18005 slave 12904
463a6c7caa958085e008ad0324213883811 0 1598793291000 2 connected
0d8799e882b249fc6bfd655ffcb5a4f89f81efe0 192.168.11.15:8006@18006 slave 1c58e
8b6ae18751ad98858aa9d7ea2c710d1511c 0 1598793291302 3 connected
12904463a6c7caa958085e008ad0324213883811 192.168.11.15:8002@18002 master - 0
1598793291607 2 connected 5461-10922
1c58e8b6ae18751ad98858aa9d7ea2c710d1511c 192.168.11.15:8003@18003 master - 0
1598793291503 3 connected 10923-16383
fb8600a4f020daeebd85a64369a9aa04b38a7d39 192.168.11.15:8001@18001 myself,slave
3d9d3cf8c6fcd7e8adc08095df9b12de7843807e 0 1598793289000 7 connected
3d9d3cf8c6fcd7e8adc08095df9b12de7843807e 192.168.11.15:8004@18004 master - 0
1598793291000 7 connected 0-5460
```

（3）增加 Redis 主节点

Redis 集群增加 Redis 主节点的命令如下。

`$ redis-cli --cluster add-node 192.168.11.15:8007 192.168.11.15:8001`

redis-cli --cluster 命令参数说明如下。

第一个参数 add-node 表示添加 Redis 节点。

第二个参数 192.168.11.15:8007 指定待增加的 Redis 主节点。

第三个参数 192.168.11.15:8001 可以是 Redis 集群中的任何一个 Redis 节点。

执行命令后，如果响应信息的最后有 "[OK] New node added correctly." 提示就代表 Redis 主节点添加成功了。

```
[root@localhost ~]# redis-cli --cluster add-node 192.168.11.15:8007 192.168.11.15:8001
>>> Adding node 192.168.11.15:8007 to cluster 192.168.11.15:8001
>>> Performing Cluster Check (using node 192.168.11.15:8001)
S: fb8600a4f020daeebd85a64369a9aa04b38a7d39 192.168.11.15:8001
   slots: (0 slots) slave
   replicates 3d9d3cf8c6fcd7e8adc08095df9b12de7843807e
S: 7bc9b1dbd1c4508828dbcb10828e20063c934c9a 192.168.11.15:8005
   slots: (0 slots) slave
   replicates 12904463a6c7caa958085e008ad0324213883811
S: 0d8799e882b249fc6bfd655ffcb5a4f89f81efe0 192.168.11.15:8006
   slots: (0 slots) slave
   replicates 1c58e8b6ae18751ad98858aa9d7ea2c710d1511c
M: 12904463a6c7caa958085e008ad0324213883811 192.168.11.15:8002
   slots:[5461-10922] (5462 slots) master
   1 additional replica(s)
M: 1c58e8b6ae18751ad98858aa9d7ea2c710d1511c 192.168.11.15:8003
   slots:[10923-16383] (5461 slots) master
```

```
   1 additional replica(s)
M: 3d9d3cf8c6fcd7e8adc08095df9b12de7843807e 192.168.11.15:8004
   slots:[0-5460] (5461 slots) master
   1 additional replica(s)
[OK] All nodes agree about slots configuration.
>>> Check for open slots...
>>> Check slots coverage...
[OK] All 16384 slots covered.
>>> Send CLUSTER MEET to node 192.168.11.15:8007 to make it join the cluster.
[OK] New node added correctly.
```

使用 cluster info 命令查看 Redis 集群信息，从图 6-29 可以看到 Redis 节点为 7 个。

图 6-29　查看 Redis 集群信息

使用 cluster nodes 命令查看 Redis 集群的节点信息。可以看到端口为 8007 的 Redis 节点状态为 master，但是现在没有散列槽分配给这个 Redis 节点。

```
192.168.11.15:8001> cluster nodes
   484b7c18a54a1690e4fda5361023d7d83d529354 192.168.11.15:8007@18007 master - 0 1598793466000 0 connected
   7bc9b1dbd1c4508828dbcb10828e20063c934c9a 192.168.11.15:8005@18005 slave 12904463a6c7caa958085e008ad0324213883811 0 1598793466000 2 connected
   0d8799e882b249fc6bfd655ffcb5a4f89f81efe0 192.168.11.15:8006@18006 slave 1c58e8b6ae18751ad98858aa9d7ea2c710d1511c 0 1598793466000 3 connected
   12904463a6c7caa958085e008ad0324213883811 192.168.11.15:8002@18002 master - 0 1598793466539 2 connected 5461-10922
   1c58e8b6ae18751ad98858aa9d7ea2c710d1511c 192.168.11.15:8003@18003 master - 0 1598793466945 3 connected 10923-16383
   fb8600a4f020daeebd85a64369a9aa04b38a7d39 192.168.11.15:8001@18001 myself,slave 3d9d3cf8c6fcd7e8adc08095df9b12de7843807e 0 1598793466000 7 connected
   3d9d3cf8c6fcd7e8adc08095df9b12de7843807e 192.168.11.15:8004@18004 master - 0 1598793466000 7 connected 0-5460
```

> 注意　新增 Redis 集群节点成功以后，新增的 Redis 节点不会有任何数据，因为它还没有分配到任何的散列槽，我们需要为新节点手动分配散列槽。

（4）为 Redis 主节点分配散列槽

为 Redis 主节点手动分配散列槽的命令如下。

```
$ redis-cli --cluster reshard 192.168.11.15:8007
```

redis-cli --cluster 命令参数说明如下。

第一个参数 reshard 表示为 Redis 节点手动分配散列槽。

第二个参数 192.168.11.15:8007 指定待分配的 Redis 主节点。

分配散列槽有以下两种方式。

方式 1：从所有主节点拿出适量的散列槽分配到目标节点，这里的目标节点指新增加的 Redis 主节点。

方式 2：从指定的主节点拿出指定数量的散列槽分配到目标节点。

针对这两种分配散列槽的方式，我们做两个实验。

实验 1：从所有的 Redis 主节点中拿出 1000 个散列槽分配给主节点 192.168.11.15:8007。

```
[root@localhost ~]# redis-cli --cluster reshard 192.168.11.15:8007
>>> Performing Cluster Check (using node 192.168.11.15:8007)
M: 484b7c18a54a1690e4fda5361023d7d83d529354 192.168.11.15:8007
   slots: (0 slots) master
S: 0d8799e882b249fc6bfd655ffcb5a4f89f81efe0 192.168.11.15:8006
   slots: (0 slots) slave
   replicates 1c58e8b6ae18751ad98858aa9d7ea2c710d1511c
S: 7bc9b1dbd1c4508828dbcb10828e20063c934c9a 192.168.11.15:8005
   slots: (0 slots) slave
   replicates 12904463a6c7caa958085e008ad0324213883811
M: 3d9d3cf8c6fcd7e8adc08095df9b12de7843807e 192.168.11.15:8004
   slots:[0-5460] (5461 slots) master
   1 additional replica(s)
S: fb8600a4f020daeebd85a64369a9aa04b38a7d39 192.168.11.15:8001
   slots: (0 slots) slave
   replicates 3d9d3cf8c6fcd7e8adc08095df9b12de7843807e
M: 1c58e8b6ae18751ad98858aa9d7ea2c710d1511c 192.168.11.15:8003
   slots:[10923-16383] (5461 slots) master
   1 additional replica(s)
M: 12904463a6c7caa958085e008ad0324213883811 192.168.11.15:8002
   slots:[5461-10922] (5462 slots) master
   1 additional replica(s)
[OK] All nodes agree about slots configuration.
>>> Check for open slots...
>>> Check slots coverage...
[OK] All 16384 slots covered.
How many slots do you want to move (from 1 to 16384)?
```

命令执行过程中，会询问要分出多少个槽。输入 1000，按 Enter 键继续。

询问分给哪个节点。输入 Redis 节点 192.168.11.15:8007 的 ID（484b7c18a54a1690e4fda5361023d7d83d529354），按 Enter 键继续。

询问从哪些主节点拿出散列槽分配到新节点。输入 all，表示从所有主节点拿出散列槽分配到主节点 192.168.11.15:8007，然后按 Enter 键继续。

```
How many slots do you want to move (from 1 to 16384)? 1000
What is the receiving node ID? 484b7c18a54a1690e4fda5361023d7d83d529354
Please enter all the source node IDs.
  Type 'all' to use all the nodes as source nodes for the hash slots.
  Type 'done' once you entered all the source nodes IDs.
Source node #1: all
```

然后在响应消息中会有提示"Do you want to proceed with the proposed reshard plan (yes/no)?"，询问是否允许这个分配散列槽的计划。输入 yes，然后按 Enter 键继续。

然后连接 Redis 集群，查看集群信息。

```
[root@localhost ~]# redis-cli -c -h 192.168.11.15 -p 8001
192.168.11.15:8001> cluster nodes
   484b7c18a54a1690e4fda5361023d7d83d529354 192.168.11.15:8007@18007 master - 0 1598793719413 8 connected 0-332 5461-5794 10923-11255
   7bc9b1dbd1c4508828dbcb10828e20063c934c9a 192.168.11.15:8005@18005 slave 12904463a6c7caa958085e008ad0324213883811 0 1598793719000 2 connected
   0d8799e882b249fc6bfd655ffcb5a4f89f81efe0 192.168.11.15:8006@18006 slave 1c58e8b6ae18751ad98858aa9d7ea2c710d1511c 0 1598793718405 3 connected
   12904463a6c7caa958085e008ad0324213883811 192.168.11.15:8002@18002 master - 0 1598793718908 2 connected 5795-10922
   1c58e8b6ae18751ad98858aa9d7ea2c710d1511c 192.168.11.15:8003@18003 master - 0 1598793718505 3 connected 11256-16383
   fb8600a4f020daeebd85a64369a9aa04b38a7d39 192.168.11.15:8001@18001 myself,slave 3d9d3cf8c6fcd7e8adc08095df9b12de7843807e 0 1598793717000 7 connected
   3d9d3cf8c6fcd7e8adc08095df9b12de7843807e 192.168.11.15:8004@18004 master - 0 1598793719514 7 connected 333-5460
```

从以上消息可以看出 Redis 节点 192.168.11.15:8007 已经有散列槽，可以在这个节点进行读写操作了，并且这个 Redis 节点是主节点。

Redis 节点 192.168.11.15:8007 的散列槽范围是 0～332，5461～5794，10923～11255，计算这个 Redis 节点分配到的槽数总和为(332 − 0 + 1)+ (5794 − 5461 + 1)+ (11255 − 10923 + 1)= 1000，即这个新增的 Redis 主节点一共分配到了 1000 个散列槽。

实验 2：从 Redis 主节点 192.168.11.15:8002 中拿出 500 个散列槽分配给 Redis 主节点 192.168.11.15:8007。

```
[root@localhost ~]# redis-cli --cluster reshard 192.168.11.15:8007
>>> Performing Cluster Check (using node 192.168.11.15:8007)
M: 484b7c18a54a1690e4fda5361023d7d83d529354 192.168.11.15:8007
   slots:[0-332],[5461-5794],[10923-11255] (1000 slots) master
S: 0d8799e882b249fc6bfd655ffcb5a4f89f81efe0 192.168.11.15:8006
   slots: (0 slots) slave
   replicates 1c58e8b6ae18751ad98858aa9d7ea2c710d1511c
S: 7bc9b1dbd1c4508828dbcb10828e20063c934c9a 192.168.11.15:8005
   slots: (0 slots) slave
   replicates 12904463a6c7caa958085e008ad0324213883811
M: 3d9d3cf8c6fcd7e8adc08095df9b12de7843807e 192.168.11.15:8004
   slots:[333-5460] (5128 slots) master
   1 additional replica(s)
S: fb8600a4f020daeebd85a64369a9aa04b38a7d39 192.168.11.15:8001
   slots: (0 slots) slave
   replicates 3d9d3cf8c6fcd7e8adc08095df9b12de7843807e
M: 1c58e8b6ae18751ad98858aa9d7ea2c710d1511c 192.168.11.15:8003
   slots:[11256-16383] (5128 slots) master
   1 additional replica(s)
M: 12904463a6c7caa958085e008ad0324213883811 192.168.11.15:8002
   slots:[5795-10922] (5128 slots) master
   1 additional replica(s)
[OK] All nodes agree about slots configuration.
>>> Check for open slots...
```

```
>>> Check slots coverage...
[OK] All 16384 slots covered.
```

命令执行过程中，会询问要分出多少个槽。输入 500，按 Enter 键继续。

询问分给哪个节点。输入 Redis 节点 192.168.11.15:8007 的 ID(484b7c18a54a1690e4fda5361023d7d83d529354)，按 Enter 键继续。

询问从哪些主节点拿出散列槽分配到新节点中。输入节点 192.168.11.15:8002 的 ID (12904463a6c7caa958085e008ad0324213883811)，表示从指定的节点 192.168.11.15:8002 拿出散列槽分配到新增节点 192.168.11.15:8007。然后输入 done，并按 Enter 键继续。

```
How many slots do you want to move (from 1 to 16384)? 500
What is the receiving node ID? 484b7c18a54a1690e4fda5361023d7d83d529354
Please enter all the source node IDs.
  Type 'all' to use all the nodes as source nodes for the hash slots.
  Type 'done' once you entered all the source nodes IDs.
Source node #1: 12904463a6c7caa958085e008ad0324213883811
Source node #2: done
```

然后在响应消息中会有提示 "Do you want to proceed with the proposed reshard plan (yes/no)?"，询问是否允许这个分配散列槽的计划。输入 yes，然后按 Enter 键继续。

然后连接 Redis 集群，查看集群信息。

```
192.168.11.15:8001> cluster nodes
484b7c18a54a1690e4fda5361023d7d83d529354 192.168.11.15:8007@18007 master - 0 1598794023000 8 connected 0-332 5461-6294 10923-11255
7bc9b1dbd1c4508828dbcb10828e20063c934c9a 192.168.11.15:8005@18005 slave 12904463a6c7caa958085e008ad0324213883811 0 1598794023747 2 connected
0d8799e882b249fc6bfd655ffcb5a4f89f81efe0 192.168.11.15:8006@18006 slave 1c58e8b6ae18751ad98858aa9d7ea2c710d1511c 0 1598794023545 3 connected
12904463a6c7caa958085e008ad0324213883811 192.168.11.15:8002@18002 master - 0 1598794023000 2 connected 6295-10922
1c58e8b6ae18751ad98858aa9d7ea2c710d1511c 192.168.11.15:8003@18003 master - 0 1598794024757 3 connected 11256-16383
fb8600a4f020daeebd85a64369a9aa04b38a7d39 192.168.11.15:8001@18001 myself,slave 3d9d3cf8c6fcd7e8adc08095df9b12de7843807e 0 1598794023000 7 connected
3d9d3cf8c6fcd7e8adc08095df9b12de7843807e 192.168.11.15:8004@18004 master - 0 1598794023043 7 connected 333-5460
```

Redis 节点 192.168.11.15:8007 的散列槽范围是 0～332，5461～6294，10923～11255，计算这个 Redis 节点分配到的槽数总和为(332 − 0 + 1)+ (6294 − 5461 + 1)+ (11255 − 10923 + 1)= 1500，即这个 Redis 主节点一共分配到了 1500 个散列槽。

至此 Redis 主节点已经添加完毕了，现在的 Redis 集群结构由三个主节点和三个从节点变成了四个主节点和三个从节点，也就是四主三从模式。

2．新增 Redis 从节点

（1）新增 Redis 节点

新增一个 Redis 节点，需要在/usr/local/redis-cluster 目录下创建相应的文件夹。本例执行以下操作创建 8008 文件夹。

```
$ cd /usr/local/redis-cluster/
$ mkdir 8008
```

复制 8001 文件夹下的 redis.conf 配置文件到 8008 文件夹下。
```
cp /usr/local/redis/conf/redis.conf /usr/local/redis-cluster/8008
```
修改 redis.conf 配置文件的配置项。
```
#    Redis
daemonize  yes

#    Redis
port  8008

#
dir  /usr/local/redis-cluster/8008/

#
cluster-enabled  yes

#
cluster-config-file  nodes-8008.conf

#
cluster-node-timeout  5000

#     Linux       IP
bind  192.168.11.15

#
protected-mode  no

#    AOF      AOF           Redis
appendonly  yes
```

（2）启动新增的 Redis 节点

启动端口为 8008 的 Redis 节点。
```
$ redis-server /usr/local/redis-cluster/8008/redis.conf
```
执行 ps -ef | grep redis 命令，从图 6-30 可以看出端口为 8008 的 Redis 节点已经启动了。

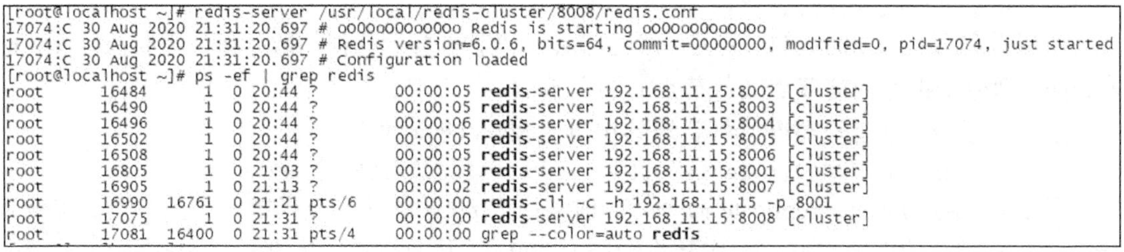

```
[root@localhost ~]# redis-server /usr/local/redis-cluster/8008/redis.conf
17074:C 30 Aug 2020 21:31:20.697 # oOOoOOooOOoo Redis is starting oOOoOOooOOoo
17074:C 30 Aug 2020 21:31:20.697 # Redis version=6.0.6, bits=64, commit=00000000, modified=0, pid=17074, just started
17074:C 30 Aug 2020 21:31:20.697 # Configuration loaded
[root@localhost ~]# ps -ef | grep redis
root     16484     1  0 20:44 ?        00:00:05 redis-server 192.168.11.15:8002 [cluster]
root     16490     1  0 20:44 ?        00:00:05 redis-server 192.168.11.15:8003 [cluster]
root     16496     1  0 20:44 ?        00:00:06 redis-server 192.168.11.15:8004 [cluster]
root     16502     1  0 20:44 ?        00:00:05 redis-server 192.168.11.15:8005 [cluster]
root     16508     1  0 20:44 ?        00:00:05 redis-server 192.168.11.15:8006 [cluster]
root     16805     1  0 21:03 ?        00:00:03 redis-server 192.168.11.15:8001 [cluster]
root     16905     1  0 21:13 ?        00:00:02 redis-server 192.168.11.15:8007 [cluster]
root     16990 16761  0 21:21 pts/6    00:00:00 redis-cli -c -h 192.168.11.15 -p 8001
root     17075     1  0 21:31 ?        00:00:00 redis-server 192.168.11.15:8008 [cluster]
root     17081 16400  0 21:31 pts/4    00:00:00 grep --color=auto redis
```

图 6-30　查看 Redis 节点进程

（3）增加 Redis 从节点

本实验配置 Redis 节点 192.168.11.15:8008 为主节点 192.168.11.15:8007 的从节点。

添加节点 192.168.11.15:8008 到 Redis 集群中去并查看集群信息，如图 6-31 所示。
```
$ redis-cli --cluster add-node 192.168.11.15:8008 192.168.11.15:8001
```

```
[root@localhost ~]# redis-cli --cluster add-node 192.168.11.15:8008 192.168.11.15:8001
>>> Adding node 192.168.11.15:8008 to cluster 192.168.11.15:8001
>>> Performing Cluster Check (using node 192.168.11.15:8001)
S: fb8600a4f020daeebd85a64369a9aa04b38a7d39 192.168.11.15:8001
   slots: (0 slots) slave
   replicates 3d9d3cf8c6fcd7e8adc08095df9b12de7843807e
M: 484b7c18a54a1690e4fda5361023d7d83d529354 192.168.11.15:8007
   slots:[0-332],[5461-6294],[10923-11255] (1500 slots) master
S: 7bc9b1dbd1c4508828dbcb10828e20063c934c9a 192.168.11.15:8005
   slots: (0 slots) slave
   replicates 12904463a6c7caa958085e008ad0324213883811
S: 0d8799e882b249fc6bfd655ffcb5a4f89f81efe0 192.168.11.15:8006
   slots: (0 slots) slave
   replicates 1c58e8b6ae18751ad98858aa9d7ea2c710d1511c
M: 12904463a6c7caa958085e008ad0324213883811 192.168.11.15:8002
   slots:[6295-10922] (4628 slots) master
   1 additional replica(s)
M: 1c58e8b6ae18751ad98858aa9d7ea2c710d1511c 192.168.11.15:8003
   slots:[11256-16383] (5128 slots) master
   1 additional replica(s)
M: 3d9d3cf8c6fcd7e8adc08095df9b12de7843807e 192.168.11.15:8004
   slots:[333-5460] (5128 slots) master
   1 additional replica(s)
[OK] All nodes agree about slots configuration.
>>> Check for open slots...
>>> Check slots coverage...
[OK] All 16384 slots covered.
>>> Send CLUSTER MEET to node 192.168.11.15:8008 to make it join the cluster.
[OK] New node added correctly.
```

图 6-31 查看 Redis 集群信息

再查看 Redis 集群的节点信息。

[root@localhost ~]# redis-cli -c -h 192.168.11.15 -p 8001

192.168.11.15:8001> cluster nodes

484b7c18a54a1690e4fda5361023d7d83d529354 192.168.11.15:8007@18007 master - 0 1598794760533 8 connected 0-332 5461-6294 10923-11255

7bc9b1dbd1c4508828dbcb10828e20063c934c9a 192.168.11.15:8005@18005 slave 12904463a6c7caa958085e008ad0324213883811 0 1598794761000 2 connected

0d8799e882b249fc6bfd655ffcb5a4f89f81efe0 192.168.11.15:8006@18006 slave 1c58e8b6ae18751ad98858aa9d7ea2c710d1511c 0 1598794761000 3 connected

12904463a6c7caa958085e008ad0324213883811 192.168.11.15:8002@18002 master - 0 1598794760634 2 connected 6295-10922

1c58e8b6ae18751ad98858aa9d7ea2c710d1511c 192.168.11.15:8003@18003 master - 0 1598794761000 3 connected 11256-16383

fb8600a4f020daeebd85a64369a9aa04b38a7d39 192.168.11.15:8001@18001 myself,slave 3d9d3cf8c6fcd7e8adc08095df9b12de7843807e 0 1598794761000 7 connected

3d9d3cf8c6fcd7e8adc08095df9b12de7843807e 192.168.11.15:8004@18004 master - 0 1598794761646 7 connected 333-5460

19bfc02f0753be1c6b14446d463e4b1440f7a850 192.168.11.15:8008@18008 master - 0 1598794761040 0 connected

可以看到端口为 8008 的节点是一个主节点，没有被分配任何的散列槽，需要执行 cluster replicate 命令来指定当前节点（从节点）的主节点。

首先，从 Redis 客户端连接到新增的端口为 8008 的节点。

[root@localhost 7007]# redis-cli -c -h 192.168.11.15 -p 8008

然后，执行 cluster replicate 命令为当前端口为 8008 的从节点指定一个主节点，也就是指定当前从节点的主节点 ID。本实验将端口为 8008 的指定为端口为 8007 的主节点的从节点，端口为 8007 的主节点的 ID 是 484b7c18a54a1690e4fda5361023d7d83d5293540 命令执行结果如图 6-32 所示。

192.168.11.15:8008> CLUSTER REPLICATE 484b7c18a54a1690e4fda5361023d7d83d529354
OK

```
192.168.11.15:8008> CLUSTER REPLICATE 484b7c18a54a1690e4fda5361023d7d83d529354
OK
```

图 6-32 将端口为 8008 的从节点指定为端口为 8007 的主节点的从节点

再查看 Redis 集群的节点信息，会发现 Redis 节点的槽范围发生了变化。

```
192.168.11.15:8008> cluster nodes
    7bc9b1dbd1c4508828dbcb10828e20063c934c9a 192.168.11.15:8005@18005 slave 12904
463a6c7caa958085e008ad0324213883811 0 1598795008000 2 connected
    fb8600a4f020daeebd85a64369a9aa04b38a7d39 192.168.11.15:8001@18001 slave 3d9d3
cf8c6fcd7e8adc08095df9b12de7843807e 0 1598795008277 7 connected
    19bfc02f0753be1c6b14446d463e4b1440f7a850 192.168.11.15:8008@18008 myself,slave
484b7c18a54a1690e4fda5361023d7d83d529354 0 1598795007000 8 connected
    3d9d3cf8c6fcd7e8adc08095df9b12de7843807e 192.168.11.15:8004@18004 master - 0
1598795007270 7 connected 333-5460
    1c58e8b6ae18751ad98858aa9d7ea2c710d1511c 192.168.11.15:8003@18003 master - 0
1598795008579 3 connected 11256-16383
    484b7c18a54a1690e4fda5361023d7d83d529354 192.168.11.15:8007@18007 master - 0
1598795008000 8 connected 0-332 5461-6294 10923-11255
    12904463a6c7caa958085e008ad0324213883811 192.168.11.15:8002@18002 master - 0
1598795008781 2 connected 6295-10922
    0d8799e882b249fc6bfd655ffcb5a4f89f81efe0 192.168.11.15:8006@18006 slave 1c58e
8b6ae18751ad98858aa9d7ea2c710d1511c 0 1598795008579 3 connected
```

至此 Redis 集群的水平扩展实验已经实现，搭建了一个四主四从的 Redis 集群，下一小节进行删除 Redis 集群节点的实验。

6.3.6　删除 Redis 集群节点

本小节进行删除 Redis 集群节点的实验，删除 Redis 集群中新增的两个 Redis 节点，即一个端口为 8007 的主节点和一个端口号为 8008 的从节点。

（1）删除从节点

使用 redis-cli --cluster del-node 命令删除 Redis 集群中的从节点，完整命令如下。

```
redis-cli --cluster del-node     ip:port     id
```

删除从节点需要指定待删除的从节点的 ID 地址和端口，以及节点 ID。待删除的从节点 192.168.11.15:8008 的 ID 是 7ef80fd1ef9a403e39d42cfefa56404f9eb6be73。删除从节点的完整命令如图 6-33 所示。

```
$ redis-cli --cluster del-node 192.168.11.15:8008 19bfc02f0753be1c6b14446d463
e4b1440f7a850
```

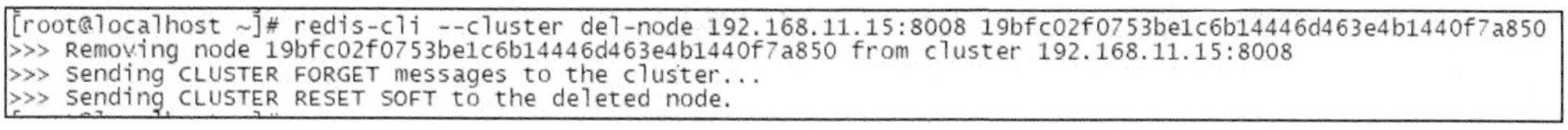

```
[root@localhost ~]# redis-cli --cluster del-node 192.168.11.15:8008 19bfc02f0753be1c6b14446d463e4b1440f7a850
>>> Removing node 19bfc02f0753be1c6b14446d463e4b1440f7a850 from cluster 192.168.11.15:8008
>>> Sending CLUSTER FORGET messages to the cluster...
>>> Sending CLUSTER RESET SOFT to the deleted node.
```

图 6-33　删除 Redis 集群的从节点

然后连接 Redis 集群，查看集群信息。

```
[root@localhost ~]# redis-cli -c -h 192.168.11.15 -p 8001
192.168.11.15:8001> cluster nodes
    484b7c18a54a1690e4fda5361023d7d83d529354 192.168.11.15:8007@18007 master - 0
1598795943590 8 connected 0-332 5461-6294 10923-11255
    7bc9b1dbd1c4508828dbcb10828e20063c934c9a 192.168.11.15:8005@18005 slave 12904
463a6c7caa958085e008ad0324213883811 0 1598795942000 2 connected
    0d8799e882b249fc6bfd655ffcb5a4f89f81efe0 192.168.11.15:8006@18006 slave 1c58e
8b6ae18751ad98858aa9d7ea2c710d1511c 0 1598795941579 3 connected
```

```
    12904463a6c7caa958085e008ad0324213883811 192.168.11.15:8002@18002 master - 0
1598795943000 2 connected 6295-10922
    1c58e8b6ae18751ad98858aa9d7ea2c710d1511c 192.168.11.15:8003@18003 master - 0
1598795942000 3 connected 11256-16383
    fb8600a4f020daeebd85a64369a9aa04b38a7d39 192.168.11.15:8001@18001 myself,slave
3d9d3cf8c6fcd7e8adc08095df9b12de7843807e 0 1598795942000 7 connected
    3d9d3cf8c6fcd7e8adc08095df9b12de7843807e 192.168.11.15:8004@18004 master - 0
1598795942985 7 connected 333-5460
```

从 Redis 集群信息可以看出从节点 192.168.11.15:8008 已经从 Redis 集群中移除，该节点的 Redis 服务也已被停止。

（2）删除主节点

删除之前增加的主节点 192.168.11.15:8007 的步骤相对麻烦一些，因为主节点已分配了散列槽，所以必须先把待删除的主节点的散列槽放入到其他可用的主节点中去，然后再进行移除节点操作，否则会出现数据丢失问题。

删除主节点 192.168.11.15:8007，要先将其散列槽分配到其他主节点上。

```
[root@localhost ~]# redis-cli --cluster reshard 192.168.11.15:8007
>>> Performing Cluster Check (using node 192.168.11.15:8007)
M: 484b7c18a54a1690e4fda5361023d7d83d529354 192.168.11.15:8007
   slots:[0-332],[5461-6294],[10923-11255] (1500 slots) master
S: 0d8799e882b249fc6bfd655ffcb5a4f89f81efe0 192.168.11.15:8006
   slots: (0 slots) slave
   replicates 1c58e8b6ae18751ad98858aa9d7ea2c710d1511c
S: 7bc9b1dbd1c4508828dbcb10828e20063c934c9a 192.168.11.15:8005
   slots: (0 slots) slave
   replicates 12904463a6c7caa958085e008ad0324213883811
M: 3d9d3cf8c6fcd7e8adc08095df9b12de7843807e 192.168.11.15:8004
   slots:[333-5460] (5128 slots) master
   1 additional replica(s)
S: fb8600a4f020daeebd85a64369a9aa04b38a7d39 192.168.11.15:8001
   slots: (0 slots) slave
   replicates 3d9d3cf8c6fcd7e8adc08095df9b12de7843807e
M: 1c58e8b6ae18751ad98858aa9d7ea2c710d1511c 192.168.11.15:8003
   slots:[11256-16383] (5128 slots) master
   1 additional replica(s)
M: 12904463a6c7caa958085e008ad0324213883811 192.168.11.15:8002
   slots:[6295-10922] (4628 slots) master
   1 additional replica(s)
[OK] All nodes agree about slots configuration.
>>> Check for open slots...
>>> Check slots coverage...
[OK] All 16384 slots covered.
How many slots do you want to move (from 1 to 16384)?
```

命令执行过程中会询问要将多少个散列槽从 Redis 集群的主节点 192.168.11.15:8007 移走。因为之前为主节点 192.168.11.15:8007 分配了 1500 个散列槽，所以我们在这里输入 1500。还要输入接收散列槽的主节点的 ID，本例中接收的主节点 192.168.11.15:8002 的 ID 是 12904463a6c7caa958085e008ad0324213883811。最后输入 all，然后按 Enter 键。

```
How many slots do you want to move (from 1 to 16384)? 1500
```

```
What is the receiving node ID? 12904463a6c7caa958085e008ad0324213883811
Please enter all the source node IDs.
  Type 'all' to use all the nodes as source nodes for the hash slots.
  Type 'done' once you entered all the source nodes IDs.
Source node #1: all
```
然后在响应消息中会有提示"Do you want to proceed with the proposed reshard plan (yes/no)?"，询问是否允许这个分配散列槽的计划。输入 yes，然后按 Enter 键继续。

给主节点 192.168.11.15:8007 分配散列槽到其他主节点后，就可以删除主节点 192.168.11.15:8007 了。

```
[root@localhost ~]# redis-cli --cluster del-node 192.168.11.15:8007 484b7c18a
54a1690e4fda5361023d7d83d529354
>>> Removing node 484b7c18a54a1690e4fda5361023d7d83d529354 from cluster 192.
168.11.15:8007
>>> Sending CLUSTER FORGET messages to the cluster...
>>> Sending CLUSTER RESET SOFT to the deleted node.
```

查看 Redis 集群信息，可以看到已经没有 192.168.11.15:8007 这个主节点了。

```
[root@localhost ~]# redis-cli -c -h 192.168.11.15 -p 8001
192.168.11.15:8001> cluster nodes
7bc9b1dbd1c4508828dbcb10828e20063c934c9a 192.168.11.15:8005@18005 slave 12904
463a6c7caa958085e008ad0324213883811 0 1598798293369 9 connected
  0d8799e882b249fc6bfd655ffcb5a4f89f81efe0 192.168.11.15:8006@18006 slave 12904
463a6c7caa958085e008ad0324213883811 0 1598798293573 9 connected
  12904463a6c7caa958085e008ad0324213883811 192.168.11.15:8002@18002 master - 0
1598798293572 9 connected 0-16383
  1c58e8b6ae18751ad98858aa9d7ea2c710d1511c 192.168.11.15:8003@18003 master - 0
1598798294378 3 connected
  fb8600a4f020daeebd85a64369a9aa04b38a7d39 192.168.11.15:8001@18001 myself,slave
12904463a6c7caa958085e008ad0324213883811 0 1598798293000 9 connected
  3d9d3cf8c6fcd7e8adc08095df9b12de7843807e 192.168.11.15:8004@18004 master - 0
1598798295388 7 connected
```

从 Redis 集群信息可以看出主节点 192.168.11.15:8007 已经从 Redis 集群中移除。

第 7 章 Redis 开发实战

本章主要介绍使用 Jedis 操作 Redis。Jedis 是 Redis 的 Java 客户端，是基于 Java 的 redis-cli，提供了完整的 Redis 命令。在开始使用 Java 操作 Redis 前，读者需要确保自己的计算机上已经安装了 Redis 服务器，且能正常使用 Redis。

7.1 搭建开发 Redis 的 Java 开发环境

本节讲解在最常见的操作系统（Windows）上安装并配置所需的 Java 开发环境，以及开发 Java 的利器 IntelliJ IDEA（简写为 IDEA）。Redis 开发安装环境信息如表 7-1 所示。

表 7-1　　　　　　　　　　　Redis 开发安装环境信息

操作系统	Windows 10　64 位
JDK	1.8.0_102
Tomcat	9.0.2
Maven	3.6.0

7.1.1 在 Windows 下安装 Java 8

Java 是由 Sun Microsystems 公司（已被 Orade 公司收购）于 1995 年 5 月推出的高级程序设计语言。Java 可运行于多个平台，例如 Windows、macOS，及其他多种 UNIX 版本的操作系统。

JDK 是 Java 的软件开发工具包，主要用于移动设备、嵌入式设备上的 Java 应用程序开发。JDK 是整个 Java 开发的核心，它包含 Java 的运行环境、Java 工具和 Java 基础的类库。使用 Java 操作 Redis，需要先按照以下步骤安装 Java。

1. 下载 JDK 8

在 Oracle 官网，找到 JDK 8 的安装包，单击"Accept License Agreement"，根据自己的操作系统下载对应的文件。本书的实例使用的操作系统为 64 位 Windows 10，故下载 Windows x64 对应的安装包 jdk-8u151-windows-x64.exe，如图 7-1 所示。

双击安装包进行安装，选择自定义安装 JDK 8，它的安装路径可以由读者自定义，笔者的安装路径放在了 D:\installed_software\jdk1_8 目录下。

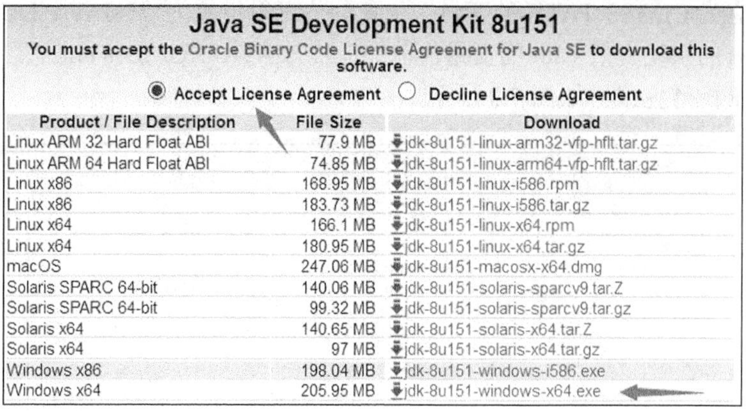

第 7 章 Redis 开发实战

图 7-1 下载 JDK 8

安装时根据提示一步步操作就可以了，注意安装路径不要使用带有中文或空格的目录，避免在之后的使用过程中出现一些莫名的错误。

2．配置 JDK 8

（1）右击"此电脑"，选择"属性"。在"系统"窗口左侧单击"高级系统设置"，打开"系统属性"对话框，如图 7-2 所示。在"系统属性"对话框的"高级"选项卡里单击"环境变量"按钮，弹出"环境变量"对话框，如图 7-3 所示。

图 7-2 从"系统属性"对话框打开"环境变量"对话框

（2）新建名为 JAVA_HOME 的变量，变量的值为之前安装的 JDK 8 的路径：D:\installed_software\jdk1_8。

143

(3)新建名为 CLASS_PATH 的变量，变量的值可以设置为"%JAVA_HOME%/lib;.;"。

(4)在已有的系统变量 Path 的变量值前加上"%JAVA_HOME%/bin;.;"。

至此 JDK 8 配置完成，如图 7-3 所示。

图 7-3　JDK 8 配置完成

下面检验 JDK 8 是否配置成功。按"Win + R"组合键，在运行框中输入 cmd 命令，如图 7-4 所示。在命令提示符窗口中执行 java -version 命令，如果出现图 7-5 所示的结果，则表明配置 JDK 8 成功了。

图 7-4　cmd 命令

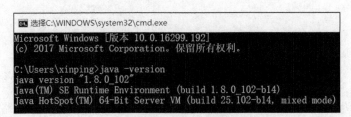

图 7-5　查看 JDK 的版本

7.1.2　安装 Tomcat 9

Tomcat 服务器是一个免费的开放源代码的 Web 应用服务器，属于轻量级应用服务器，在中小型操作系统和并发访问用户不是很多的场合下被普遍使用，是开发和调试 JSP 程序的首选。

读者可以根据自己的操作系统，去 Apache 官网选择合适的版本下载，笔者下载的是 Windows 64 位版本的 apache-tomcat-9.0.24.zip。将压缩包解压缩后放到指定目录就可以使用，无须安装。Tomcat 9（以下简称 Tomcat）解压缩后的目录如图 7-6 所示。

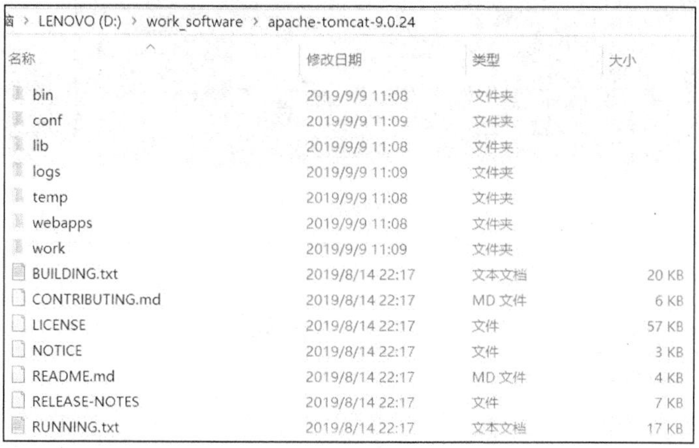

图 7-6　Tomcat 解压缩后的目录

Tomcat 关键目录及文件作用如下。

- bin：用于存放各种平台下启动和关闭 Tomcat 的脚本文件。在该目录中有两个非常关键的文件 startup.bat 和 shutdown.bat，前者是 Windows 下启动 Tomcat 的文件，后者是对应的关闭文件。
- conf：Tomcat 的各种配置文件，其中 server.xml 为服务器的主配置文件，web.xml 为所有 Web 应用的配置文件，tomcat-users.xml 用于定义 Tomcat 的用户信息、配置用户的权限与安全。
- lib：此目录存放 Tomcat 和所有 Web 应用都能访问的 JAR 包。
- logs：用于存放 Tomcat 的日志文件，Tomcat 的所有日志文件都存放在此目录中。
- temp：临时文件夹，Tomcat 运行时如果有临时文件将保存于此目录中。
- webapps：Web 应用的发布目录，把 Java Web 站点或 WAR 文件放入这个目录中，就可以通过 Tomcat 访问了。
- work：Tomcat 解析 JSP 所生成的 Servlet 文件放在这个目录中。

启动 Tomcat，在 Tomcat 的 bin 目录下找到 startup.bat 文件并双击运行。此时会弹出一个黑色窗口的控制台，如图 7-7 所示。黑色窗口的控制台不要关闭，如果关闭就相当于把 Tomcat 关闭了。

在浏览器地址栏中输入 http://localhost:8080 或者 http://127.0.0.1:8080，如果看到图 7-8 所示界面，则证明 Tomcat 启动成功了。Tomcat 默认的访问端口是 8080。

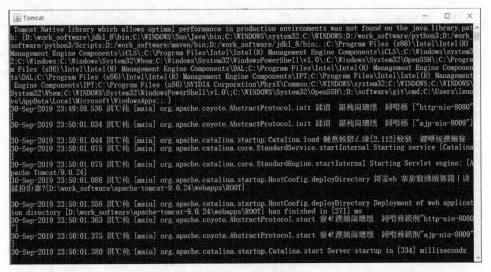

图 7-7　在 Windows 下启动 Tomcat

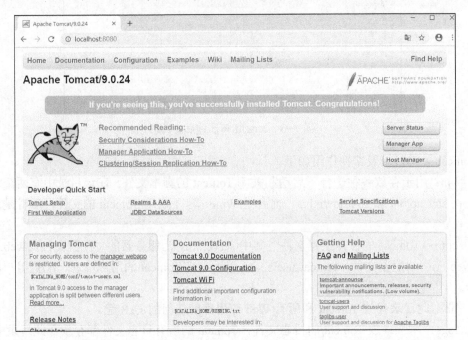

图 7-8　启动 Tomcat

7.1.3　搭建 IntelliJ IDEA 开发环境

俗话说，"工欲善其事，必先利其器"。在做 Java 开发之前，一个好的 IDEA 会使编程效率得到大幅度提高。IntelliJ IDEA 是目前最流行的 Java 开发环境之一。本小节主要介绍 IntelliJ IDEA 开发环境的安装和配置。IntelliJ IDEA 不但可以开发 Java 应用，还可以作为 Java 源代码的阅读器。

1. 安装 IntelliJ IDEA

IntelliJ IDEA 是用于 Java 开发的集成环境（也可用于其他语言）。IntelliJ IDEA 在业界被公认为最好的 Java 开发环境之一，尤其在智能代码助手、代码自动提示、重构、Java EE 支持、Ant、JUnit、CVS 整合、代码审查、创新的 GUI 设计等方面的功能可以说是超常的。

IntelliJ IDEA 官网上对不同的操作系统（Windows、macOS、Linux）都有两个版本可供下载。Ultimate 为旗舰版，功能全面，插件丰富，但是按年收费，功能上类似于 MyEclipse。

Community 为社区版，免费使用，功能相对而言不是很丰富，但是不影响开发使用，功能上类似于 Eclipse。本书选用 IntelliJ IDEA 社区版作为 Java 开发环境。IntelliJ IDEA 社区版的安装很简单，按照提示一步步在 Windows 上操作就可以安装成功。

IntelliJ IDEA 下载页面如图 7-9 所示。

图 7-9　IntelliJ IDEA 下载页面

2．配置 IntelliJ IDEA

因为本书篇幅所限，具体的 IntelliJ IDEA 配置 JDK、Maven 和 Tomcat 部分请参考 11.3 节配置 IntelliJ IDEA。

7.2　使用 Java 操作 Redis

7.2.1　连接 Redis 的两种方式

使用 IDEA 新建 Maven 项目。Redis 官方推荐的 Java 版客户端是 Jedis，Jedis 支持命令、事务、管道。我们对 Redis 数据的操作，都可以通过客户端 Jedis 来完成。可在 Jedis 官网找到其参考文档。本实例名为"Redis\Chapter 07\JedisDemo 1"。

使用 IDEA 新建 Maven 项目，需要在 Maven 项目中导入 pom.xml 文件。

```
<dependency>
    <groupId>redis.clients</groupId>
```

```
        <artifactId>jedis</artifactId>
        <version>3.1.0</version>
</dependency>
```

导入的 pom.xml 文件实际引入了如下 JAR 包，如图 7-10 所示。

文件名	日期	类型	大小
commons-pool2-2.6.2.jar	2019/9/4 17:21	JAR 文件	127 KB
jedis-3.1.0.jar	2019/9/4 17:21	JAR 文件	632 KB
slf4j-api-1.7.25.jar	2018/11/19 14:13	JAR 文件	41 KB

图 7-10　Jedis 客户端需要的 JAR 包

1. Jedis 直连

Jedis 直连相当于一个 TCP 连接，数据传输完成后关闭连接，如图 7-11 所示。

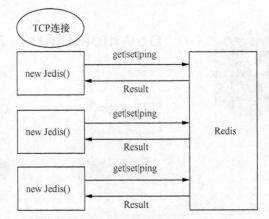

1.生成Jedis对象；2.Jedis执行命令；3.返回执行结果；4.关闭Jedis连接。

图 7-11　Jedis 直连

本实例使用"RedisDemo.java"，内容如下。

```java
package com.dxtd.demo;

import redis.clients.jedis.Jedis;

public class RedisDemo {

    public static void main(String[] args){
        //生成一个Jedis对象，这个对象负责和指定的Redis节点进行通信
        Jedis jedis = new Jedis("127.0.0.1", 6379);
        //jedis执行set命令
        jedis.set("name","xinping");
        //jedis执行get命令
        String name = jedis.get("name");
        System.out.println(name);
        jedis.close();
    }
}
```

该程序的输出如下。
xinping
打开一个终端,并使用 redis-cli 命令连接到 Redis 服务器。
```
127.0.0.1:6379> KEYS *
1) "name"
127.0.0.1:6379> GET name
"xinping"
```
可以看到数据已经保存到 Redis 服务器上了。

2．Jedis 连接池连接

使用 Jedis 连接池可以不用创建新的 Jedis 对象连接 Redis 服务器,这大大减少创建和回收 Redis 连接的开销,如图 7-12 所示。

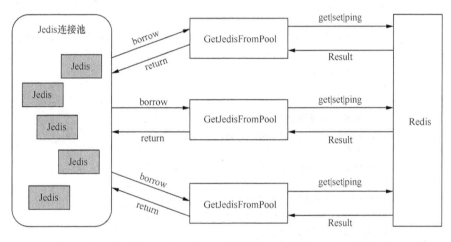

1.从连接池借Jedis对象;2.Jedis执行命令;3.返回执行结果;4.归还Jedis对象给连接池。

图 7-12　Jedis 连接池连接

表 7-2 所示是使用 Jedis 直连和 Jedis 连接池连接这两种方案的对比。

表 7-2　　　　使用 Jedis 直连和 Jedis 连接池连接这两种方案的对比

	优点	缺点
Jedis 直连	(1)简单、方便地创建 Jedis 对象。 (2)适用于连接数比较少且使用时间较长,可以构成长连接的场景	(1)存在每次新建和关闭 TCP 的开销。 (2)每次都会创建 Jedis 对象,系统资源可能无法在有效范围内进行控制,会存在连接容易发生泄漏的问题。 (3)Jedis 对象本身是线程不安全的
Jedis 连接池连接	(1)Jedis 对象是预先生成的,不需要随用随创建,随用完随关闭,降低了开销。 (2)连接池这种形式能够更好地保护和控制资源使用,有固定的参数去控制最大连接数或者空闲数目等	相对 Jedis 直连来说,使用起来更麻烦,特别是资源管理上需要多个参数来保证,一旦出现规划不合理的情况就会出现问题,例如连接池满了、大量连接空闲、连接超时等情况

本实例使用"JedisPoolDemo.java",内容如下。
```
package com.dxtd.demo;
```

```java
import redis.clients.jedis.Jedis;
import redis.clients.jedis.JedisPool;
import redis.clients.jedis.JedisPoolConfig;

public class JedisPoolDemo {

    public static void main(String[] args){
        //创建连接池配置对象
        JedisPoolConfig jpc = new JedisPoolConfig();
        //设置最大闲置个数
        jpc.setMaxIdle(30);
        //设置最小闲置个数
        jpc.setMinIdle(10);
        //设置最大连接数
        jpc.setMaxTotal(50);
        //创建连接池对象。host 为连接 Redis 主机 IP 地址；port 为 Redis 的默认端口
        JedisPool jedisPool = new JedisPool(jpc, "127.0.0.1", 6379);
        Jedis resource = null;
        try{

            //获取连接资源
            resource = jedisPool.getResource();
            //放值
            resource.set("name", "hello redis");
            //取值
            String name = resource.get("name");
            System.out.println("name=" + name);
        }catch(Exception e){
            e.printStackTrace();
        }finally {
            if (resource != null) {
                //这里使用的 close 不代表关闭连接，指的是归还资源
                jedisPool.close();
            }
        }
    }
}
```

该程序输出如下。

name=hello redis

7.2.2 操作 String

本小节使用 Jedis 操作 String。在本例中使用 Jedis 的 set()方法添加数据，使用 append()方法拼接数据，使用 del()方法删除某个 key，使用 mset()方法设置多个 key-value 键值对，使用 incr()方法对 key 进行加 1 操作。

本实例文件名为 Redis\Chapter07\JedisDemo1\src\main\java\com\dxtd\demo\RedisDemo2.java，内容如下。

```java
public static void testString() {
```

```java
        // 连接Redis服务器,127.0.0.1:6379
        Jedis jedis = new Jedis("127.0.0.1", 6379);
        // -----添加数据----------
        jedis.set("name", "wang");
        System.out.println(jedis.get("name"));// 执行结果: wang

        jedis.append("name", " is my friend"); // 拼接
        System.out.println(jedis.get("name"));

        jedis.del("name"); // 删除某个key
        System.out.println(jedis.get("name"));
        // 设置多个key-value键值对
        jedis.mset("name", "xinping", "age", "35", "qq", "759949947");
        jedis.incr("age"); // 进行加1操作
        System.out.println(jedis.get("name") + "-" + jedis.get("age") + "-" + jedis.get("qq"));
        jedis.close();
    }
```

该程序输出如下。

```
wang
wang is my friend
null
xinping-36-759949947
```

7.2.3 操作 Map

本小节使用 Jedis 操作 Map。在本例中使用 Jedis 的 hmset() 方法添加多个数据,使用 hmget() 方法得到多个 key 的值,使用 hdel() 方法删除某个 key,使用 hkeys() 方法返回 Map 中的所有 key,使用 hvals() 方法返回 Map 中的所有 value。

本实例使用 "RedisDemo2.java",内容如下。

```java
public static void testMap() {
        // -----添加数据----------
        Map<String, String> map = new HashMap<String, String>();
        map.put("name", "wang");
        map.put("age", "22");
        map.put("qq", "123456");

        // 连接Redis服务器,127.0.0.1:6379
        Jedis jedis = new Jedis("127.0.0.1", 6379);
        jedis.hmset("user", map);
        // 取出user中的name,执行结果:[minxr]。注意,结果是一个泛型的List
        // 第一个参数是存入Redis中Map的key,后面跟的是放入Map中的key,后面的key可以
        // 跟多个,是可变参数
        List<String> rsmap = jedis.hmget("user", "name", "age", "qq");

        // 删除Map中的某个key
        jedis.hdel("user", "age");
        System.out.println(jedis.hmget("user", "age")); // 因为删除了,所以返回的是null
        System.out.println(jedis.hlen("user")); // 返回key为user的键中存放的值的个数2
        System.out.println(jedis.exists("user"));// 是否存在key为user的记录,存在
```

则返回 true

```
        System.out.println(jedis.hkeys("user"));// 返回 Map 中的所有 key
        System.out.println(jedis.hvals("user"));// 返回 Map 中的所有 value

        Iterator<String> iter = jedis.hkeys("user").iterator();
        while (iter.hasNext()) {
            String key = iter.next();
            System.out.println(key + ":" + jedis.hmget("user", key));
        }
        jedis.close();
    }
```

该程序输出如下。

```
[null]
2
true
[name, qq]
[wang, 123456]
name:[wang]
qq:[123456]
```

7.2.4 操作 List

本小节使用 Jedis 操作 List。在本例中使用 Jedis 的 lpush()方法向 key 添加列表数据，使用 lrange()方法遍历列表中的数据。

本实例使用 "RedisDemo2.java"，内容如下。

```java
public static void testList() {
        // 连接 Redis 服务器, 127.0.0.1:6379
        Jedis jedis = new Jedis("127.0.0.1", 6379);
        // 开始操作前, 先删除 key 为 "java framework" 的所有内容
        jedis.del("java framework");
        System.out.println(jedis.lrange("java framework", 0, -1));
        // 先向 key 为 "java framework" 的列表按照从左到右的顺序存放 3 条数据
        jedis.lpush("java framework", "spring");
        jedis.lpush("java framework", "struts");
        jedis.lpush("java framework", "hibernate");
        // jedis.lrange()返回列表中指定区间内的元素，其中 0 表示列表的第一个元素，-1 表示列表的最后一个元素
        System.out.println(jedis.lrange("java framework", 0, -1));
        // 开始操作前, 先移除 key 为 "java framework" 的所有内容
        jedis.del("java framework");
        // 再向 key 为"java framework"的列表尾部（最右边）存放三条数据
        jedis.rpush("java framework", "spring");
        jedis.rpush("java framework", "struts");
        jedis.rpush("java framework", "hibernate");
        System.out.println(jedis.lrange("java framework", 0, -1));
        jedis.close();
    }
```

该程序输出如下。

```
[]
[hibernate, struts, spring]
```

[spring, struts, hibernate]

7.2.5 操作 Set

本小节使用 Jedis 操作 Set。在本例中使用 Jedis 的 sadd()方法添加元素，使用 smembers() 方法获取所有加入的 value，使用 sismember()方法判断集合中的元素是否存在。

本实例使用 "RedisDemo2.java"，内容如下。

```java
public static void testSet() {
    // 连接 Redis 服务器，127.0.0.1:6379
    Jedis jedis = new Jedis("127.0.0.1", 6379);
    // 添加元素
    jedis.sadd("usernames", "lisi","wangwu","xinping","zhangsan");
    // 获取所有加入的 value
    System.out.println(jedis.smembers("usernames"));
    // 判断 who 是不是 user 集合的元素
    System.out.println(jedis.sismember("usernames", "who"));
    System.out.println(jedis.srandmember("usernames"));
    // 返回集合的元素个数
    System.out.println(jedis.scard("usernames"));
    jedis.close();
}
```

该程序输出如下。

```
[zhangsan, lisi, wangwu, xinping]
false
xinping
4
```

7.2.6 排序

本小节使用 Jedis 对数据进行排序。在本例中使用 Jedis 的 rpush()方法、lpush 方法向列表添加数据，然后使用 sort()方法进行排序，最后使用 lrange()方法显示排序后的列表。

本实例使用 "RedisDemo2.java"，内容如下。

```java
public static void testSort()   {
        // 连接 Redis 服务器，127.0.0.1:6379
        Jedis jedis = new Jedis("127.0.0.1", 6379);

        // Jedis 排序
        // 注意，此处的 rpush()和 lpush()是对 List 的操作，是一个双向链表
        jedis.del("a");// 先清除数据，再加入数据进行测试
        jedis.rpush("a", "1");
        jedis.lpush("a", "6");
        jedis.lpush("a", "3");
        jedis.lpush("a", "9");
        System.out.println(jedis.lrange("a", 0, -1));// [9, 3, 6, 1]
        System.out.println(jedis.sort("a"));  //输出排序后的结果
        System.out.println(jedis.lrange("a", 0, -1));
}
```

该程序输出如下。

```
[9, 3, 6, 1]
```

```
[1, 3, 6, 9]
[9, 3, 6, 1]
```

7.2.7 Redis 存储图片

本小节使用 Redis 来存储一张图片。首先从网络下载一张图片，然后使用 Base64 对图片进行编码，再使用 Base64 对图片进行解码，最后将图片保存到本地 E 盘根目录下。

在百度图片中搜索熊猫，在搜索结果中可以看到一张可爱的大熊猫图片，如图 7-13 所示。我们要把这张图片存储在 Redis 中。

图 7-13 大熊猫图片

1. 使用 Base64 对图片进行编码和解码

本实例使用"Base64ImageUtils.java"，内容如下。

```java
package com.dxtd.demo;

import java.awt.image.BufferedImage;
import java.io.ByteArrayOutputStream;
import java.io.File;
import java.io.FileOutputStream;
import java.io.IOException;
import java.net.MalformedURLException;
import java.net.URL;
import javax.imageio.ImageIO;
import sun.misc.BASE64Decoder;
import sun.misc.BASE64Encoder;

public class Base64ImageUtils {

    /**
     * 对网络图片进行 Base64 编码
     *
     * @param imageUrl
     *            图片的 URL 路径，例如 http://www.aaa.com/b.jpg
     * @return
```

```java
         */
        public static String encodeImgageToBase64(URL imageUrl) {// 将图片文件转化为
字节数组字符串, 并对其进行 Base64 编码处理
            ByteArrayOutputStream outputStream = null;
            try {
                BufferedImage bufferedImage = ImageIO.read(imageUrl);
                outputStream = new ByteArrayOutputStream();
                ImageIO.write(bufferedImage, "jpg", outputStream);
            } catch (MalformedURLException e1) {
                e1.printStackTrace();
            } catch (IOException e) {
                e.printStackTrace();
            }
            // 对字节数组字符串进行 Base64 编码
            BASE64Encoder encoder = new BASE64Encoder();
            return encoder.encode(outputStream.toByteArray());// 返回 Base64 编码过的
字节数组字符串
        }

        /**
         * 将本地图片进行 Base64 编码
         *
         * @param imageFile
         *              图片的 URL 路径, 例如 F:/.....xx.jpg
         * @return
         */
        public static String encodeImgageToBase64(File imageFile) {// 将图片文件转化
为字节数组字符串, 并对其进行 Base64 编码处理
            ByteArrayOutputStream outputStream = null;
            try {
                BufferedImage bufferedImage = ImageIO.read(imageFile);
                outputStream = new ByteArrayOutputStream();
                ImageIO.write(bufferedImage, "jpg", outputStream);
            } catch (MalformedURLException e1) {
                e1.printStackTrace();
            } catch (IOException e) {
                e.printStackTrace();
            }
            // 对字节数组字符串进行 Base64 编码
            BASE64Encoder encoder = new BASE64Encoder();
            return encoder.encode(outputStream.toByteArray());// 返回 Base64 编码过的
字节数组字符串
        }

        /**
         * 对 Base64 编码过的图片进行解码, 并保存到指定目录
         *
         * @param base64
         *              Base64 编码的图片信息
         * @return
         */
```

```java
public static void decodeBase64ToImage(String base64, String path,
                                String imgName) {
    BASE64Decoder decoder = new BASE64Decoder();
    try {
        FileOutputStream write = new FileOutputStream(new File(path
                + imgName));
        byte[] decoderBytes = decoder.decodeBuffer(base64);
        write.write(decoderBytes);
        write.close();
    } catch (IOException e) {
        e.printStackTrace();
    }
}

public static void main(String [] args){
    URL url = null;
    try {
        url = new URL("https://graph.baidu.com/thumb/691874602,911901597.jpg");
    } catch (MalformedURLException e) {
        e.printStackTrace();
    }
    String encoderStr = Base64ImageUtils.encodeImgageToBase64(url);
    System.out.println(encoderStr);

    Base64ImageUtils.decodeBase64ToImage(encoderStr, "E:/", "pandas.jpg");
}
```

在控制台查看 Base64 编码后的图片信息，如图 7-14 所示。

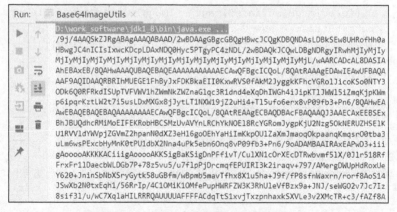

图 7-14 Base64 编码后的图片信息

本程序将网络上的大熊猫图片保存为本地磁盘 E 下的 pandas.jpg。下载的图片如图 7-15 所示，和网络上的图片是一样的。

图片转化成 String 需要注意以下两点。

- 我们可以在 Redis 中存储图片的 Base64 编码或者解码，以 key-value 键值对格式存储，key 为普通字符串，value 为图片的 Base64 编码，获取 key 后再使用 Base64 解码就可以了。

- 我们也可以在 Redis 中存储图片的网络 URL 地址或者本地的 Path 路径。图片可以存储到磁盘中，然后在 Redis 中以图片的网络 URL 地址或者本地的 Path 路径作为 value 进行存储。

2. 图片转化成 Object

要将图片转化成 Object，可直接在 Redis 中存储图片，使用 Java 的序列化/反序列化机制进行处理。本实例使用 "ImageRedisTest.java"，内容如下。

图 7-15 下载的图片

```java
package com.dxtd.demo;

import java.net.MalformedURLException;
import java.net.URL;
import redis.clients.jedis.Jedis;


public class ImageRedisTest {

    private static Jedis jedis = null;

    /**
     * 初始化 Jedis 对象
     *
     * @throws Exception
     */
    public ImageRedisTest() {
        jedis = new Jedis("127.0.0.1", 6379);
    }

    /**
     * 将图片写入 Redis 中
     *
     */
    public void saveImage() {
       URL url = null;
       try {
          url = new URL("https://graph.baidu.com/thumb/691874602,911901597.jpg");
       } catch (MalformedURLException e) {
          e.printStackTrace();
       }
       String encoderStr = Base64ImageUtils.encodeImgageToBase64(url);
       System.out.println(encoderStr);

       jedis.set("image:e:/pandas.jpg".getBytes(), encoderStr.getBytes());
    }

    /**
     * 从 Redis 中获得图片，保存到本地
```

```java
     *
     */
    public void getImage() {
        System.out.println(jedis.get(("image:e:/pandas.jpg")));
        String encoderStr = jedis.get("image:e:/pandas.jpg");
        Base64ImageUtils.decodeBase64ToImage(encoderStr, "E:/", "pandas2.jpg");

    }

    public static void main(String[] args) {
        ImageRedisTest test = new ImageRedisTest();
        test.saveImage();
        //test.getImage();
    }
}
```

运行程序，使用 redis-cli 命令连接 Redis 服务器，可以看到在 Redis 数据库里已经存在 key 为 image:e:/pandas.jpg 了。获取这个 key 的值，如图 7-16 所示，可以看到已经成功把图片存储到 Redis 了。

图 7-16 成功把图片存储到 Redis

7.2.8 Redis 存储 Object

Redis 存储 Object（对象）的时候，要进行对象的序列化与反序列化操作。本实例使用"Person.java"，保存用户数据，内容如下：

```java
package com.dxtd.demo;

import java.io.Serializable;

public class Person  implements Serializable {
    private int id;
    private String name;

    public Person() {
    }

    public Person(int id, String name) {
        super();
```

```java
        this.id = id;
        this.name = name;
    }

    public int getId() {
        return id;
    }

    public void setId(int id) {
        this.id = id;
    }

    public String getName() {
        return name;
    }

    public void setName(String name) {
        this.name = name;
    }

}
```
创建序列化工具，本实例使用"SerializeUtil.java"，内容如下。
```java
package com.dxtd.demo;

import java.io.ByteArrayInputStream;
import java.io.ByteArrayOutputStream;
import java.io.ObjectInputStream;
import java.io.ObjectOutputStream;

public class SerializeUtil {
    /**
     * 序列化
     *
     * @param object
     * @return
     */
    public static byte[] serialize(Object object) {
        ObjectOutputStream oos = null;
        ByteArrayOutputStream baos = null;
        try {
            baos = new ByteArrayOutputStream();
            oos = new ObjectOutputStream(baos);
            oos.writeObject(object);
            byte[] bytes = baos.toByteArray();
            return bytes;
        } catch (Exception e) {
            e.printStackTrace();
        }
        return null;
    }
```

```java
/**
 * 反序列化
 *
 * @param bytes
 * @return
 */
public static Object unserialize(byte[] bytes) {
    ByteArrayInputStream bais = null;
    try {
        bais = new ByteArrayInputStream(bytes);
        ObjectInputStream ois = new ObjectInputStream(bais);
        return ois.readObject();
    } catch (Exception e) {
        e.printStackTrace();
    }
    return null;
}
```

使用 Redis 存储 Object 对象，本实例使用 "PersonRedisTest.java"，内容如下。

```java
package com.dxtd.demo;

import redis.clients.jedis.Jedis;

public class PersonRedisTest {

    private static Jedis jedis = null;

    /**
     * 初始化 Jedis 对象
     *
     * @throws Exception
     */
    public PersonRedisTest() {
        jedis = new Jedis("127.0.0.1", 6379);
    }

    /**
     * 序列化写对象，将 Person 对象写入 Redis 中
     *
     */
    public void setObject() {
        jedis.set("person:100".getBytes(),
            SerializeUtil.serialize(new Person(100, "zhangsan")));
        jedis.set("person:101".getBytes(),
            SerializeUtil.serialize(new Person(101, "xinping")));
    }

    /**
     * 反序列化取对象，用 Jedis 获取对象
     *
```

```
    */
    public void getObject() {
      byte[] data100 = jedis.get(("person:100").getBytes());
      Person person100 = (Person) SerializeUtil.unserialize(data100);
      System.out.println(String.format("person:100->id=%s,name=%s",
             person100.getId(), person100.getName()));

      byte[] data101 = jedis.get(("person:101").getBytes());
      Person person101 = (Person) SerializeUtil.unserialize(data101);
      System.out.println(String.format("person:101->id=%s,name=%s",
             person101.getId(), person101.getName()));
    }

    public static void main(String[] args) {
      PersonRedisTest rt = new PersonRedisTest();
      rt.setObject();
      rt.getObject();
    }
}
```

该程序输出如下。

```
person:100->id=100,name=zhangsan
person:101->id=101,name=xinping
```

查看 Redis 存储的对象，如图 7-17 所示，可以看出已经把 Java 对象存储到 Redis 了。

```
127.0.0.1:6379> GET person:100
"\xac\xed\x00\x05sr\x00\x14com.dxtd.demo.Person>}\xff\x05\xe0\n\xf1\x02\x00\x02I\x00\x02idL\x00\x04namet\x00\x12Ljava/lang/S
tring;xp\x00\x00\x00dt\x00\x0bzhangsan"
127.0.0.1:6379> GET person:101
"\xac\xed\x00\x05sr\x00\x14com.dxtd.demo.Person>}\xff\x05\xe0\n\xf1\x02\x00\x02I\x00\x02idL\x00\x04namet\x00\x12Ljava/lang/S
tring;xp\x00\x00\x00et\x00\axinping"
```

图 7-17　Redis 存储的对象

7.2.9　Redis 存储和计算用户访问量

有这样一个场景，我们需要统计在 2020 年 8 月 6 日 22 点到 23 点之间系统某个页面的访客数，指的是某个用户不管访问这个页面几次都算作访问页面一次是不重复计数的。key 设置为 user:login:2020080622。这种场景下适合使用 HyperLogLog 来存储和计算用户访问量，HyperLogLog 使用统计概率的算法，牺牲数据的精准性来节省内存的占用空间。

本实例使用"RedisPFCountTest.java"，内容如下。

```
public static void testHyperLogLog(){
    Jedis jedis = new Jedis("127.0.0.1",6379);
    // 模拟在 2020 年 8 月 6 日 22 点，存储 100 万条用户访问数据
    for (int i = 0; i < 100 * 10000; i++) {
        jedis.pfadd("user:login:2020080622", "user-"+ i );
    }

    long total = jedis.pfcount("user:login:2020080622");
    System.out.printf("key=user:login:202008062200 count=%d", total);

    jedis.close();
}
```

运行程序,输出如下。

```
key=user:login:2020080622 count=1003993
```

从返回结果可以看出,使用 HyperLogLog 存储 100 万条用户访问数据,计算用户访问量多出了 3993 条,存在一定的误差。

然后再通过 redis-rdb-tools 统计 user:login:2020080622 这个 key 的信息,发现它只占用 14400Byte 的硬盘空间,也就是 14KB。

```
$ rdb -c memory /usr/local/redis/bin/dump.rdb
database,type,key,size_in_bytes,encoding,num_elements,len_largest_element,expiry
0,string,user:login:2020080622,14400,string,12304,12304,
```

所以 HyperLogLog 适合统计用户访问量、日活跃用户量、月活跃用户量此类对精确度要求不高的场景。

7.3 Redis 调用方式

本节介绍 Redis 的 4 种常用调用方式。本节实例使用 "RedisTransDemo.java",本节内容使用的代码都保存在这个文件内。

7.3.1 普通同步

普通同步是一种最简单和最基础的调用方式,对于简单的数据存取需求,可以通过这种方式调用。

```java
public static void jedisNormal() {
    Jedis jedis = new Jedis("localhost");
    long start = System.currentTimeMillis();
    for (int i = 0; i < 10; i++) {
        String result = jedis.set("n:" + i,   ""+i);
    }
    long end = System.currentTimeMillis();
    System.out.println("Simple SET: " + ((end - start)/1000.0) + " seconds");
    jedis.disconnect();
}
```

该程序输出如下。

```
Simple SET: 0.031 seconds
```

7.3.2 事务

Redis 事务可以一次执行多个命令,有以下两个特性。

- 隔离性:事务的所有命令都会被序列化,按顺序执行,事务执行完后才会执行其他客户端的命令。
- 原子性:事务中的命令要么全部被执行,要么全部不执行。

```java
public static void jedisTrans() {
    Jedis jedis = new Jedis("localhost");
    long start = System.currentTimeMillis();
    // 开启事务
    Transaction tx = jedis.multi();
    for (int i = 0; i < 100; i++) {
```

```
            tx.set("t:" + i, i + "");
        }

        // 提交事务
        List<Object> results = tx.exec();
        long end = System.currentTimeMillis();
        System.out.println("Transaction SET: " + ((end - start)/1000.0) + " seconds");
        jedis.disconnect();
    }
```
该程序输出如下。
```
Transaction SET: 0.047 seconds
```

7.3.3 管道

管道是一种两个进程之间进行单向通信的机制。在 Redis 中有时候我们需要采用异步的方式，一次发送多个命令，并且不同步等待其返回结果，这样可以取得非常好的执行效率。

```
public static void jedisPipelined() {
        Jedis jedis = new Jedis("localhost");
        Pipeline pipeline = jedis.pipelined();
        long start = System.currentTimeMillis();
        for (int i = 0; i < 100; i++) {
            pipeline.set("p:" + i, "" + i);
        }
        List<Object> results = pipeline.syncAndReturnAll();
        long end = System.currentTimeMillis();
        System.out.println("Pipelined SET: " + ((end - start)/1000.0) + " seconds");
        jedis.disconnect();
    }
```
该程序输出如下。
```
Pipelined SET: 0.051 seconds
```

7.3.4 管道中调用事务

有时候我们需要异步执行命令，但是又希望多个命令是连续的，所以我们就采用管道中调用事务的方式。Jedis 支持在管道中调用事务。

```
public static void jedisCombPipelineTrans() {
    Jedis jedis = new Jedis("localhost");
    long start = System.currentTimeMillis();
    Pipeline pipeline = jedis.pipelined();
    pipeline.multi();
    for (int i = 0; i < 100000; i++) {
        pipeline.set("" + i, "" + i);
    }
    pipeline.exec();
    List<Object> results = pipeline.syncAndReturnAll();
    long end = System.currentTimeMillis();
    System.out.println("Pipelined transaction: " + ((end - start) / 1000.0) + " seconds");
    jedis.disconnect();
}
```

该程序输出如下。

```
Pipelined transaction: 0.463 seconds
```

7.4 Redis 集群与 Java

本节使用 Jedis 连接一个 Redis 集群，一个 Redis 集群环境由 6 个节点组成，每个节点有不同的端口，有 3 个主节点和 3 个从节点。Redis 集群搭建请参考 6.3.2 小节开始 Redis 集群搭建。

本实例使用 "RedisClusterDemo.java"，内容如下。

```java
public static void testJedisCluster()   {
    JedisPoolConfig config = new JedisPoolConfig();
    config = new JedisPoolConfig();
    config.setMaxTotal(60000);        // 设置最大连接数
    config.setMaxIdle(1000);          // 设置最大空闲数
    config.setMaxWaitMillis(3000);    // 设置超时时间
    config.setTestOnBorrow(true);

    HashSet<HostAndPort> nodes = new HashSet<>();
    nodes.add(new HostAndPort("192.168.11.15", 8001));
    nodes.add(new HostAndPort("192.168.11.15", 8002));
    nodes.add(new HostAndPort("192.168.11.15", 8003));
    nodes.add(new HostAndPort("192.168.11.15", 8004));
    nodes.add(new HostAndPort("192.168.11.15", 8005));
    nodes.add(new HostAndPort("192.168.11.15", 8006));
    JedisCluster cluster = new JedisCluster(nodes, config);
    cluster.set("book", "redis");
    System.out.println("集群测试  key=book,value=" + cluster.get("book"));
    cluster.close();
}
```

程序运行返回以下结果。

```
集群测试   key=book , value=redis
```

使用 redis-cli -c -h 192.168.11.15 -p 8001 连接 Redis 集群，可以看到数据已经成功插入，如图 7-18 所示。

```
[root@localhost ~]# redis-cli -c -h 192.168.11.15 -p 8001
192.168.11.15:8001> Get book
"redis"
```

图 7-18 查看插入 Redis 集群的数据

7.5 实例 1：使用 Redis 获取用户的共同好友

可以使用 Redis 来存储社交关系，例如有这样一个需求：在微博中 zhangsan 有一批好友，lisi 有另外一批好友，现在需要查询 zhangsan 和 lisi 的共同好友。为了实现这个需求，可以使用 Redis 的 Set 数据类型。Set 类型是 String 类型的无序集合，集合中的元素是唯一的，这表示集合中不能出现重复的元素。可以使用 Redis 的 SINTER key 命令获得给定集合的交集。

7.5.1 初始化数据

首先，使用 redis-cli 命令连接 Redis 服务器，执行以下命令。

```
# 清空当前数据库
127.0.0.1:6379> FLUSHDB
OK

# 添加用户 zhangsan 拥有的好友集合
127.0.0.1:6379> SADD zhangsan "friend_1"
(integer) 1
127.0.0.1:6379> SADD zhangsan "friend_2"
(integer) 1
127.0.0.1:6379> SADD zhangsan "friend_3"
(integer) 1

# 添加用户 lisi 拥有的好友集合
127.0.0.1:6379> SADD lisi "friend_1"
(integer) 1
127.0.0.1:6379> SADD lisi "friend_3"
(integer) 1
127.0.0.1:6379> SADD lisi "friend_5"
(integer) 1

# 显示用户 zhangsan 的好友集合
127.0.0.1:6379> SMEMBERS zhangsan
1) "friend_2"
2) "friend_1"
3) "friend_3"

# 显示用户 lisi 的好友集合
127.0.0.1:6379> SMEMBERS lisi
1) "friend_5"
2) "friend_1"
3) "friend_3"

# 获取 zhangsan 和 lisi 的共同好友
127.0.0.1:6379> SINTER zhangsan lisi
1) "friend_1"
2) "friend_3"
```

7.5.2 使用 Jedis 获取用户的共同好友

使用 Jedis 完成同样的操作，即获取 zhangsan 和 lisi 的共同好友。本实例使用 "TestSNS.java"，内容如下。

```
package com.redis;

import redis.clients.jedis.Jedis;

public class TestSNS {
```

```java
public static void main(String[] args) {
    Jedis jedis = new Jedis("127.0.0.1", 6379);
    //清空当前数据库
    jedis.flushDB();

    //zhangsan 的好友
    jedis.sadd("zhangsan", "friend_1");
    jedis.sadd("zhangsan", "friend_2");
    jedis.sadd("zhangsan", "friend_3");
    System.out.println("zhangsan 的好友 =>"+ jedis.smembers("zhangsan"));

    //lisi 的好友
    jedis.sadd("lisi", "friend_1");
    jedis.sadd("lisi", "friend_3");
    jedis.sadd("lisi", "friend_5");
    System.out.println("lisi 的好友 =>"+ jedis.smembers("lisi"));

    //获取 zhangsan 和 lisi 的共同好友
    System.out.println("zhangsan和lisi的共同好友 =>"+ jedis.sinter("zhangsan","lisi"));
    //释放资源
    jedis.close();
  }
}
```

运行程序得到结果，和在 Redis 中的统计结果是一样的。

```
zhangsan 的好友 =>[friend_3, friend_2, friend_1]
lisi 的好友 =>[friend_3, friend_1, friend_5]
zhangsan 和 lisi 的共同好友 =>[friend_3, friend_1]
```

7.6 实例 2：在 Tomcat 上使用 Redis 保存 Session

当用户访问量大和应用服务器使用集群来部署时，使用 Tomcat 默认自带的 Session 就不能满足需求了。当然解决方法有很多，本节提供了一个解决方法，就是使用 Redis 来保存 Session。好处就是使用 Session 的代码没有任何变化，Tomcat 默认把 Session 保存到 Redis 中，把 Redis 缓存作为一个 Session 的存储系统。我们先看一下分布式 Session 的概念。

7.6.1 分布式 Session

单服务器 Web 应用中，Session 信息只存在该服务器中，这是前几年的流行方式。但是近几年随着分布式系统的流行，单系统已经不能满足日益增长的百万级用户的需求了，集群方式部署服务器已在很多公司运用起来。当高并发量的请求到达服务器的时候通过负载均衡的方式发送到集群中的某个服务器，这样就可能导致同一个用户的多次请求被发送到集群的不同服务器上，就会出现取不到 Session 数据的情况，于是 Session 的共享就成了一个问题。

如图 7-19 所示，假设用户包含登录信息的 Session 都记录在 Web-Server1 上，反向代理如果将请求路由到 Web-Server2 上，就找不到相关登录信息，从而导致用户需要重新登录。

分布式系统 Session 一致性的问题，可以采用 Session 复制（同步）来解决，持久化 Session

到 Redis 中，这样每个 Web-Server 都包含全部的 Session。Web-Server 可以支持 Web 应用原有的功能，不需要修改代码，如图 7-20 所示。

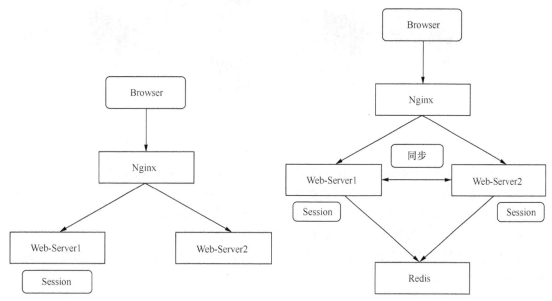

图 7-19　分布式系统 Session 一致性的问题　　图 7-20　Session 复制，持久化 Session 到 Redis

7.6.2　持久化 Tomcat Session 到 Redis

HTTP 是无状态协议，这意味着每次客户端访问网页时，都要单独打开一个服务器，因此服务器不会记录先前客户端请求的任何信息。

Session 表示客户端与服务器的一次会话。从客户端打开浏览器并连接到服务器开始，到客户端关闭浏览器离开这个服务器结束，被称为一个会话。当一个客户端访问一个服务器时，可能会在服务器的几个页面之间切换，服务器可以通过 Session 来保存客户端的用户信息，如果用户关闭了浏览器，那么 Session 就会丢失。改进方法是把 Redis 作为 Tomcat 服务器的缓存，保存 Session。

实验环境基于 CentOS 7，实验环境需要的 Nginx、Tomcat 和 JDK 的安装请参考 7.6.3 小节的内容。实验环境如表 7-3 所示。

表 7-3　　　　在 Tomcat 上使用 Redis 作为缓存服务器的实验环境

名称	IP 地址	端口	版本	操作系统
Nginx	192.168.11.14	80	Nginx 1.17.4	CentOS 7
Tomcat-1	192.168.11.14	8080	Tomcat 7	
Tomcat-2	192.168.11.15	8080	Tomcat 7	
Redis	192.168.11.14	6379	Redis 6.0.6	
JDK	192.168.11.14		JDK 13	
JDK	192.168.11.15		JDK 13	

Redis 作为缓存服务器的实验拓扑如图 7-21 所示。

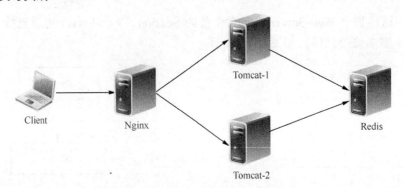

图 7-21　Redis 作为缓存服务器的实验拓扑

这里搭建一个有两台 Tomcat 服务器的小集群。在图 7-21 中，Nginx 作为反向代理，实现动静分离的功能，将客户端动态请求根据权重（Weight）随机分配给两台 Tomcat 服务器，Redis 作为两台 Tomcat 服务器的共享 Session 数据服务器。为了保证网络的连通性，把两台 Linux 主机的防火墙全部关闭，具体内容请参考 11.1.1 小节关闭防火墙。

Redis 在 IP 地址为 192.168.11.14 的 Linux 服务器上的安装请参考 1.2.2 小节在 Linux 下安装 Redis。

7.6.3　安装服务器 Tomcat 和反向代理服务器 Nginx

1．安装服务器 Tomcat

首先在两台计算机上安装 JDK 13，并设置系统变量。具体操作如下。

（1）在 Linux 上安装 Java。

首先下载 JDK 13。在 Oracle 官网找到 JDK 13 的下载地址。在下载页面中单击"Accept License Agreement"，并根据自己的操作系统选择对应的版本。本小节以 CentOS 7 64 位操作系统为例，下载 Linux 版本的 Java 压缩包 jdk-13_linux-x64_bin.tar.gz，如图 7-22 所示。

图 7-22　下载 JDK 13

在 /usr/local 目录下创建 java 文件夹。

```
$ mkdir -p /usr/local/java
```

把 jdk-13_linux-x64_bin.tar.gz 上传到 CentOS 7 服务器。手动解压缩 JDK 13 的压缩包 jdk-13_linux-x64_bin.tar.gz，然后设置系统变量。

```
$ tar -zxvf jdk-13_linux-x64_bin.tar.gz
```
复制解压缩的 JDK 13 到/usr/local/java 目录下。
```
$ mv jdk-13 /usr/local/java
```
使用 vi 命令修改/etc/profile 配置文件,设置系统变量。
```
$ vi /etc/profile
```
在/etc/profile 配置文件中添加如下内容。
```
export JAVA_HOME=/usr/local/java
export PATH=$JAVA_HOME/bin:..:$PATH:
export CLASS_PATH=$JAVA_HOME/lib:..:
```
让修改的/etc/profile 配置文件生效。
```
$ source /etc/profile
```

注意

/usr/local/java 是笔者在本机上安装 JDK 13 的位置,在 Linux 上安装的所有软件都推荐部署在/usr/local 目录下。

在 Linux 下路径之间使用冒号(:)分隔。

验证 JDK 13 有效性。
```
[root@localhost upload]# java -version
java version "13" 2019-09-17
Java(TM) SE Runtime Environment (build 13+33)
Java HotSpot(TM) 64-Bit Server VM (build 13+33, mixed mode, sharing)
```
(2)安装 Tomcat 7。

在 Apache 官网下载 Tomcat 7 的 Linux 版本压缩包 apache-tomcat-7.0.96.tar.gz,如图 7-23 所示。

图 7-23 下载 Tomcat 7 的 Linux 版本压缩包

然后解压缩 Tomcat 7 的压缩包并授予执行权限。
```
$ tar -zxvf apache-tomcat-7.0.96.tar.gz
$ chmod -R 777 apache-tomcat-7.0.96
```
启动 Tomcat 7,进入解压缩后的 Tomcat 7 目录。

```
$ cd apache-tomcat-7.0.96/bin
$ ./startup.sh
```

访问 http://127.0.0.1:8080，如果看到图 7-24 所示页面就说明启动 Tomcat 7 成功了。

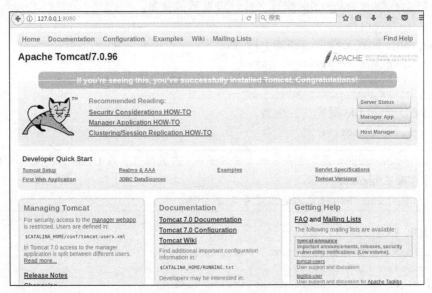

图 7-24　启动 Tomcat 7

停止 Tomcat 7 可以使用如下命令。

```
$ ./shutdown.sh
```

2. 安装反向代理服务器 Nginx

Nginx 是一个高性能的 HTTP 和反向代理 Web 服务器，同时也提供了 IMAP/POP3/SMTP 服务。在 Nginx 官网下载最新的 Nginx 的 Linux 版本压缩包 nginx-1.17.4.tar.gz，如图 7-25 所示。

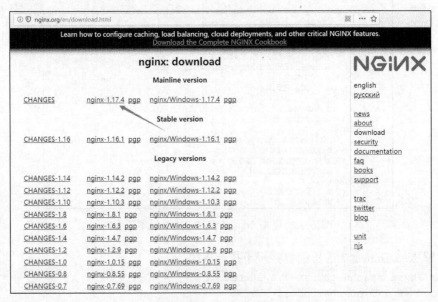

图 7-25　下载 Nginx

要做到负载均衡，就需要使用代理服务器将接收的请求均衡地分发到各服务器中。这里的代理服务器就是 Nginx，分发请求的服务器使用的是 Tomcat。

在主机 192.168.11.14 上安装 Nginx，具体步骤如下。

（1）解压缩 Nginx 压缩包。

```
$ tar -zxvf nginx-1.17.4.tar.gz
```

在安装 Nginx 之前先检查是否已安装 Nginx 的一些模块依赖的 lib 库，例如 g++、gcc、pcre-devel、openssl-devel 和 zlib-devel。

```
$ yum -y install gcc-c++  pcre pcre-devel  zlib zlib-devel openssl openssl-devel --setopt=protected_multilib=false
```

（2）编译 Nginx 源代码。

把 Nginx 安装到/usr/local/nginx 目录下。

```
$ cd nginx-1.17.4
$ ./configure --prefix==/usr/local/nginx
```

编译 Nginx。

```
$ make
```

安装 Nginx。

```
$ make install
```

安装 Nginx 成功后，查看 Nginx 的安装目录，即/usr/local/nginx 目录，如图 7-26 所示。

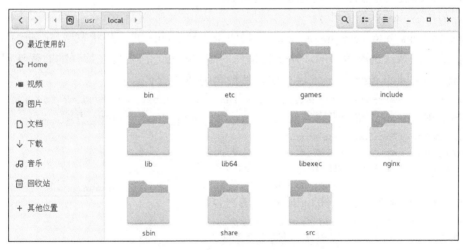

图 7-26　查看 Nginx 的安装目录

安装 Nginx 后，查看 Nginx 版本。

```
[root@localhost local]# /usr/local/nginx/sbin/nginx -v
nginx version: nginx/1.17.4
```

（3）启动 Nginx。

```
$ /usr/local/nginx/sbin/nginx
```

然后在浏览器中访问 http://127.0.0.1 查看 Nginx 是否启动成功。Nginx 默认占用的端口是 80，HTTP 默认使用的端口也是 80，所以访问 http://127.0.0.1:80 和 http://127.0.0.1 的效果是一样的，如图 7-27 所示。

如果看到图 7-27 所示页面就说明 Nginx 启动成功了。

图 7-27 启动 Nginx

（4）配置 Nginx 服务器。

关闭 Nginx 服务器。

```
$ /usr/local/nginx/sbin/nginx -s stop
```

其中参数 -s 表示强制关闭 Nginx 服务器。

测试 Nginx 的配置文件是否正确。

```
$ /usr/local/nginx/sbin/nginx -t
```

重启 Nginx，修改配置文件后，使用以下命令重新加载，使配置文件生效。

```
$ /usr/local/nginx/sbin/nginx -s reload
```

（5）启用 Nginx 状态信息。

可以通过 ngx_http_stub_status_module 模块来监控 Nginx 的一些状态信息。在编译 Nginx 的时候，此模块默认是不编译的。从源码编译安装 Nginx 时，需要在编译的时候使用如下命令添加 ngx_http_stub_status_module 模块，然后再编译安装 Nginx。

```
$ ./configure --prefix=/usr/local/nginx --with-http_stub_status_module
```

然后使用 nginx -V 命令来查看是否有 ngx_http_stub_status_module 模块。

```
$ /usr/local/nginx/sbin/nginx -V
nginx version: nginx/1.17.4
built by gcc 4.8.5 20150623 (Red Hat 4.8.5-39) (GCC)
configure arguments: --prefix=/usr/local/nginx --with-http_stub_status_module
```

然后修改 nginx 的配置文件 nginx.conf 的内容。

```
$ vi /usr/local/nginx/conf/nginx.conf
```

添加以下内容到 server 节点里，然后保存文件，退出到命令行。

```
location ~ ^/nginx-status {
    allow 127.0.0.1;
    deny all;
    stub_status on;
    access_log off;
}
```

修改配置后，使用如下命令重新加载 Nginx 的配置文件，使配置文件生效。

```
$ /usr/local/nginx/sbin/nginx -s reload
```

重启 Nginx 后，在浏览器中访问 http://127.0.0.1/nginx-status 进入 Nginx 状态信息页面，如图 7-28 所示。

页面中参数含义如下。

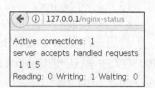

图 7-28 Nginx 的状态信息页面

- Active connections：表示 Nginx 的当前的活跃连接数。
- server：表示 Nginx 共处理的连接数。
- accepts：表示 Nginx 已接受的连接数。
- handled requests：表示 Nginx 从启动到现在处理的客户端请求总数。
- Reading：表示 Nginx 读取到客户端的 Header 信息数。
- Writing：表示 Nginx 返回给客户端的 Header 信息数。
- Waiting：表示 Nginx 已经处理完，正在等候下一次请求指令的驻留连接（开启 keep-alive 设置的情况下，这个值等于 Active connections – (Reading + Writing))。

7.6.4 配置 Tomcat 集群

1．部署 Web 项目

Session 保存在服务器中，Cookie 保存在用户浏览器本地或内存中。用户在发起一个 HTTP 请求时，在请求头中会有 Cookie，在 Cookie 中有一个 Session ID 值（Tomcat 中称为 JSESSIONID），通过这个值获取到服务器对应的 Session，如图 7-29 所示。

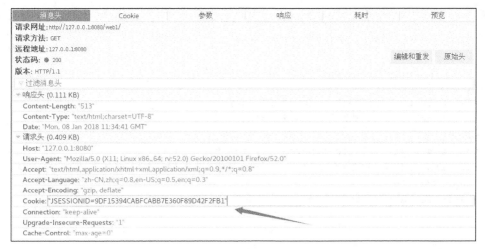

图 7-29 查看 Cookie 值

用户在给 Session 赋值时，我们可以为用户生成一个唯一的 Cookie 值作为 Session ID 存储在用户的客户端，将该 Cookie 值作为缓存的 key 和 Session 值一起存入 Redis 缓存中。

使用 IntelliJ IDEA 新建 Java EE Web 项目，本实例文件名为"Redis\Chapter 07\web1"，如图 7-30 所示。

修改 index.jsp 文件，具体内容如下。

```
<%@ page language="java" contentType="text/html;charset=UTF-8"%>
<html>
    <head><title>TomcatA</title></head>
```

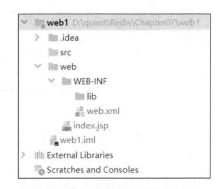

图 7-30 使用 IntelliJ IDEA 新建 Java EE Web 项目

```
    <body>

      <table align="centre" border="1">
        <tr>
          <td>Session ID</td>
          <td><%= session.getId() %></td>
        </tr>
        <tr>
          <td>Created on</td>
          <td><%= session.getCreationTime() %></td>
        </tr>
      </table>
   </body>
</html>
sessionID:<%=session.getId()%>
<br>
SessionIP:<%=request.getServerName()%>
<br>
SessionPort:<%=request.getServerPort()%>
<br>
<%
//为了区分,第二个页面显示的内容可以是 This is Tomcat Server 222
out.println("This is Tomcat Server 111");

Object obj = session.getAttribute("counter");
if( null == obj){
 session.setAttribute("counter" , new Integer(1));
 out.println("该页面被访问了 1 次<br/>");
}else{
 int counterVal = Integer.parseInt( obj.toString() );
 out.println("该页面被访问了 " + counterVal +" 次");
 counterVal ++;
 session.setAttribute("counter", counterVal);
}

%>
```

把 web1 项目部署到主机 192.168.11.15 的 Tomcat 的%/tomcat7/webapps 目录下,然后稍微修改 index.jsp 文件上的内容以示区别,以便区分请求被分发到了哪个 Tomcat 上。修改 index.jsp 文件的内容如下。

```
<%@ page language="java" contentType="text/html;charset=UTF-8"%>
<html>
  <head><title>TomcatA</title></head>
  <body>

    <table align="centre" border="1">
      <tr>
        <td>Session ID</td>
        <td><%= session.getId() %></td>
      </tr>
      <tr>
```

```html
            <td>Created on</td>
            <td><%= session.getCreationTime() %></td>
        </tr>
    </table>
  </body>
</html>
sessionID:<%=session.getId()%>
<br>
SessionIP:<%=request.getServerName()%>
<br>
SessionPort:<%=request.getServerPort()%>
<br>
<%
out.println("This is Tomcat Server 222");

Object obj = session.getAttribute("counter");
if( null == obj){
 session.setAttribute("counter" , new Integer(1));
 out.println("该页面被访问了 1 次<br/>");
}else{
 int counterVal = Integer.parseInt( obj.toString() );
 out.println("该页面被访问了 " + counterVal +" 次");
 counterVal ++;
 session.setAttribute("counter", counterVal);
}

%>
```

把 web1 项目分别部署到两台主机的 Tomcat 服务器上，然后访问 http://localhost/web1 会发现 Session ID 不一致，如果要保持 Session ID 一致，需要配置 Tomcat 集群。

把 web1 项目部署在主机 192.168.11.14 的%/tomcat7/webapps/目录下。访问 http://192.168.11.14:8080/web1，查看页面中的 Session ID，如图 7-31 所示。

把 web1 项目部署在主机 192.168.11.15 的%/tomcat7/webapps/目录下。访问 http://192.168.11.15:8080/web1，查看另一个页面中的 Session ID，如图 7-32 所示。

图 7-31　查看页面中的 Session ID

图 7-32　查看另一个页面中的 Session ID

为了配置 Tomcat 集群，需要配置 Nginx 服务器。

2．配置 Nginx 服务器

修改主机 192.168.11.14 上的 Nginx 配置文件 nginx.conf。

```
vi /usr/local/nginx/conf/nginx.conf
```
修改 Nginx 配置文件 nginx.conf 后，对 nginx.conf 配置文件内容进行精简，精简后的配置文件内容如图 7-33 所示。

在 nginx.conf 配置文件里配置 Nginx 反向代理服务器使用的主要命令如图 7-33 所示。

（1）使用 upstream 命令配置后端服务器集群。

（2）使用 proxy_pass 命令配置需要转发的路径地址。

```
worker_processes  1;

events {
    worker_connections  1024;
}

http {
    include       mime.types;
    default_type  application/octet-stream;

    sendfile        on;

    keepalive_timeout  65;

    # 服务器集群
    upstream tomcat {   # # 服务器集群的名称
        # 服务器配置，权重越大，Web应用获得分配请求的概率越大
        server 192.168.11.14:8080 weight=1;
        server 192.168.11.15:8080 weight=1;
    }

    server {
        listen       80;
        server_name  localhost;

        location / {
            root   html;
            index  index.html index.htm;
            proxy_pass http://tomcat;
        }
        error_page   500 502 503 504  /50x.html;
        location = /50x.html {
            root   html;
        }
    }
}
```

图 7-33　nginx.conf 配置文件

在本例中将 web1 项目部署在两台 Tomcat 服务器上，IP 地址分别为 192.168.11.14 和 192.168.11.15，反向代理服务器 Nginx 部署在 IP 地址为 192.168.11.14 的计算机上，proxy_pass 的 URL 配置为 http://tomcat。如果在 IP 地址为 192.168.11.14 的计算机上访问 http://127.0.0.1，Nginx 会把请求地址转向 http://tomcat。Tomcat 是 Nginx 配置的服务器集群名称，Nginx 会根据权重把请求分配给名称为 Tomcat 的服务器集群上对应的 http://192.168.11.14:8080 和 http://192.168.11.15:8080 的 Web 应用上，权重越大，Web 应用获得分配请求的概率越大。

然后重新启动 Nginx。

```
$ /usr/local/nginx/sbin/nginx -s reload
```

在主机 192.168.11.14 上访问 http://localhost/ web1，会发现一个问题：每次刷新页面都会得到不同的 Session ID 值，如图 7-34 所示。我们还需要修改 Tomcat 的配置文件，使用 Redis 保存 Session ID 值。

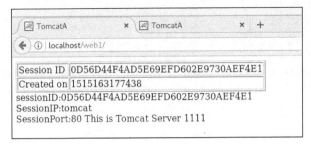

图 7-34　刷新页面会得到不同的 Session ID 值

7.6.5　配置 Tomcat 使用 Redis 管理 Session

首先，在两台主机上修改 conf/context.xml 文件的内容，重新启动 Tomcat。

（1）在两台主机的 Tomcat 的 conf/context.xml 文件里都增加如下内容：在实验环境中第一台测试服务器的 IP 地址为 192.168.11.14，也就是 Redis 服务器所在主机的 IP 地址。

```xml
<?xml version='1.0' encoding='utf-8'?>

<Context>
    <WatchedResource>WEB-INF/web.xml</WatchedResource>
<Valve className="com.orangefunction.tomcat.redissessions.RedisSessionHandlerValve" />
  <Manager className="com.orangefunction.tomcat.redissessions.RedisSessionManager"
      host="192.168.11.14"
      port="6379"
      database="0"
      />
</Context>
```

参数说明如下。
- host：对外提供服务的主机。
- port：对外提供服务的端口。

下载最新的 Jedis 模块、Tomcat Redis Session Manager 模块和 Apache Commons Pool 模块，把三个模块的 jedis-3.1.0.jar、tomcat-redis-session-manager1.2.jar、commons-pool2-2.0.jar 复制到%/Tomcat7/lib 目录下。

（2）修改 redis.conf 配置文件，修改 bind 对应的值为如下内容。

```
bind 127.0.0.1 192.168.11.14
```

然后重新启动 Redis 服务器。

```
./redis-server redis.conf
```

客户端连接 Redis 服务器。

```
redis-cli -h 192.168.11.14 -p 6379
```

在 IP 地址为 192.168.11.14 的 Linux 主机上再次访问 http://localhost:8080/web1，会发现 Session ID 值不变，如图 7-35 所示。

可以看出，分别访问了不同的 Tomcat，但是得到的 Session ID 值却是相同的，说明达到了集群的目的，这样就实现了负载均衡。假设服务器在运行过程中，其中一个 Tomcat 崩溃了，仍然还有另一个 Tomcat 可以正常访问服务。

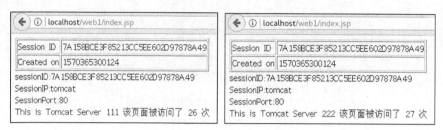

图 7-35 Session ID 值不变

访问 Redis 会发现多了一个 key，说明在 Tomcat 上使用 Redis 保存了 Session，如图 7-36 所示。

```
192.168.11.14:6379> keys *
1) "7A158BCE3F85213CC5EE602D97878A49"
192.168.11.14:6379> get 7A158BCE3F85213CC5EE602D97878A49
"\xac\xed\x00\x05sr\x00Dcom.orangefunction.tomcat.redissessions.SessionSerializationMetadataB\xd9\xd9\xf7v\xa2\xdbL\x03\x00\x01[\x00\x15sessionAttribu
tesHasht\x00\x02[Bxpw\x14\x00\x00\x00\x10Ho}k\xb9\xbf\x84^9\x97\x9doe,B?xsr\x00\x0ejava.lang.Long;\x8b\xe4\x90\xcc\x8f#\xdf\x02\x00\x01J\x00\x05valuex
r\x00\x10java.lang.Number\x86\xac\x95\x1d\x0b\x94\xe0\x8b\x02\x00\x00xp\x00\x00\x01m\xa1\x10\xdd\x9csq\x00~\x00\x03\x00\x00\x01m\xa1\x10\xdd\x9csr\x00
\x11java.lang.Integer\x12\xe2\xa0\xa4\xf7\x81\x878\x02\x00\x01I\x00\x05valuexq\x00~\x00\x04\x00\x1bw@sr\x00\x11java.lang.Boolean\xcd r\x80\xd5\x9c\xfa
\xee\x02\x00\x01Z\x00\x05valuexp\x00sq\x00~\x00\t\x01sq\x00~\x00\x03\x00\x00\x01m\xa1\x10\xf3\xe3t\x00 7A158BCE3F85213CC5EE602D97878A49sq\x00~\x00\a\x
00\x00\x00\x01t\x00\acountersq\x00~\x00\a\x00\x00\x00\x1aw\b\x00\x00\x01m\xa1\x10\xdd\x9c"
```

图 7-36 访问 Redis 服务器

第 8 章 Spring Boot 与 Redis 整合应用

本章讲解 Spring Boot 与 Redis 的整合应用。Spring Boot 是企业中常用的微服务架构，掌握 Spring Boot 与 Redis 的整合可以提高我们工作中的开发效率。

8.1 Spring Boot 项目搭建与 Redis 整合应用

8.1.1 Spring Boot 简介

微服务是一种架构风格，是一种使用一套微小服务来开发单个应用的方法。每个服务运行在自己的进程中，通过轻量的通信机制联系，经常是基于 HTTP 资源的 API。这些服务基于业务能力构建，能够通过自动化部署方式独立部署。微服务，简单地说就是将一个大服务（项目）划分为多个子服务（项目）。典型微服务架构如图 8-1 所示。

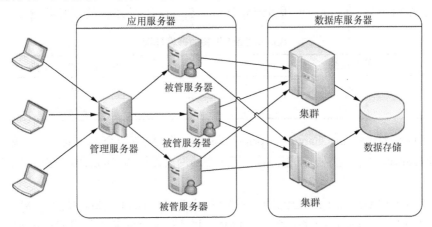

图 8-1 典型微服务架构

从图 8-1 可以看出微服务的好处：每个微服务组件都是简单、灵活的，能够独立部署；微服务应用不需要一个庞大的应用服务器来支撑；微服务之间是松耦合的，内部是高内聚的，每个微服务很容易按需扩展；微服务架构与语言、工具无关，可以自由选择合适的语言和工具，高效地完成业务目标。

Spring Boot 是 Spring 的一套快速配置"脚手架"，可以基于 Spring Boot 快速开发单个微服务。Spring Boot 本身并不提供 Spring 的核心特性以及扩展功能，只是用于快速、敏捷地开发新一代基于 Spring 的应用程序，如图 8-2 所示。也就是说，Spring Boot 并不是用来替代 Spring 的解决方案，而是和 Spring 紧密结合用于提升 Spring 开发者体验的工具。

Spring Boot 在开发过程中大量使用"约定优先配置"的思想来摆脱 Spring 中各类繁复纷杂的配置。采用 Spring Boot 可以大大地简化开发模式，同时它集成了大量常用的第三方库配置（例如 Redis、MongoDB、JPA、RabbitMQ、Quartz 等），Spring Boot 中这些第三方库几乎可以零配置地开箱即用。大部分的 Spring Boot 应用都只需要非常少量的配置代码，这使开发者能够更加专注于业务逻辑。

图 8-2　Spring Boot 与 Spring 的关系

Spring Boot 不等于微服务，它只是一套开源框架，跟 SSM（Spring+Spring MVC+MyBatis）差不多，只是基于 Spring Boot 来开发微服务相当方便，所以这两个词一般都是成对出现的。当我们的服务越来越多时，就可以通过 Spring Cloud 来统一管理这些服务了。Spring Cloud 才算是真正的微服务架构，而 Spring Boot 是 Java EE 一站式解决方案。使用 Spring Boot 可以利用微服务开发小型的 Web 项目。

Spring Boot 官网上的文档质量很高，如果想快速理解 Spring Boot，浏览官网文档是最快的途径之一。

8.1.2　使用 Spring Initializr 新建项目

本小节介绍在 Windows 下使用 Spring Boot。Spring Boot 依赖的开发环境如表 8-1 所示。

表 8-1　Spring Boot 依赖的开发环境

操作系统	Windows 10　64 位
JDK	1.8
Maven	3.6.0

本实例目录为"Redis\Chapter08\springbootDemo"。创建一个 Spring Boot 项目，可以使用 Spring Initializr 方式。这种方式很简单，通过在线网页设置项目的选项，即可自动生成 Spring Boot 项目。

首先在 https://start.spring.io/ 上创建一个简单的 Spring Boot 项目，类型为 Maven Project，开发语言为 Java，Spring Boot 版本为 2.2.0 M6，并填写项目的基本信息 Group 和 Artifact，如图 8-3 所示。最后单击"Generate the Project"按钮下载 springbootDemo.zip 压缩包。

本书主要以 IntelliJ IDEA 进行实验，把 springbootDemo.zip 解压缩到本地硬盘，然后使用 IntelliJ IDEA 导入工程，选择"File"→"Open"，选择 springbootDemo 在本地硬盘的路径。如果是第一次配置 Spring Boot，可能需要等待 IntelliJ IDEA 下载相应的依赖包。默认创建好的项目结构如图 8-4 所示。本实例文件名为"Redis\Chapter08\springbootDemo"。

springbootDemo 的项目结构很清晰，相比 Java EE 项目少了很多配置文件，默认生成的文件主要有 4 个。

第 8 章 Spring Boot 与 Redis 整合应用

- SpringbootDemoApplication.java：一个带有 main()方法的类，用于启动应用程序。
- SpringbootDemoApplicationTests.java：一个空的 JUnit 测试类，它加载了一个使用 Spring Boot 字典配置功能的 Spring 应用程序上下文。
- application.properties：一个空的 properties 文件，可以根据需要添加配置属性。
- pom.xml：Maven 构建项目的说明文件。

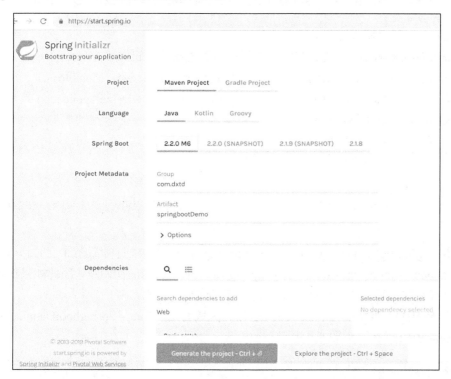

图 8-3　创建 Spring Boot 项目

图 8-4　创建好的项目结构

传统基于 Java 的 Web 应用需要配置 web.xml 文件，将 Web 应用打包成 JAR 文件放入应用服务器（Tomcat、WebLogic 等）中运行。如果基于 Spring Boot 开发微服务，将变得更加简单，只需要在 pom.xml 文件中引入 spring-boot-starter-web 开发依赖模块。

```
<dependency>
    <groupId>org.springframework.boot</groupId>
    <artifactId>spring-boot-starter-web</artifactId>
</dependency>
```

pom.xml 文件中默认有两个模块。

- spring-boot-starter：核心模块，包括自动配置支持、日志和 YAML。
- spring-boot-starter-test：测试模块，包括 JUnit、Hamcrest、Mockito。

本实例使用"SpringbootDemoApplication.java"，SpringbootDemoApplication 是 Spring Boot 项目的入口类，它的关键源代码如下。

```
@SpringBootApplication
public class SpringbootDemoApplication {
    public static void main(String[] args)
    {
        SpringApplication.run(SpringbootDemoApplication.class, args);
    }
}
```

其中，@SpringBootApplication 注解是 Spring Boot 启动项目时需要加上的。需要说明的是，SpringbootDemoApplication 是整个 Spring Boot 项目启动的初始点，因此如果在项目启动时需要进行某些资源初始化处理，那么最好都在该类中完成。

双击 SpringbootDemoApplication，在代码区域右击 "Run 'SpringbootDemoApplic…'" 启动 Spring Boot 项目，如图 8-5 所示。

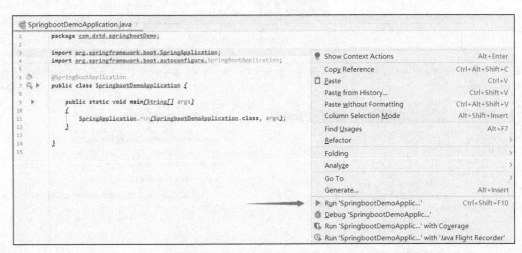

图 8-5　启动 Spring Boot 项目

我们可以看到 IntelliJ IDEA 控制台输出 Spring Boot 的启动信息，如图 8-6 所示。

如图 8-6 所示，可以看出 Tomcat 服务器默认开启了 8080 端口。要访问这个项目提供的服务，可以在浏览器的地址栏中输入 http://localhost:8080/。我们在项目中没有手动地去配置 Tomcat 服务器，因为 Spring Boot 内置了 Tomcat 服务器。

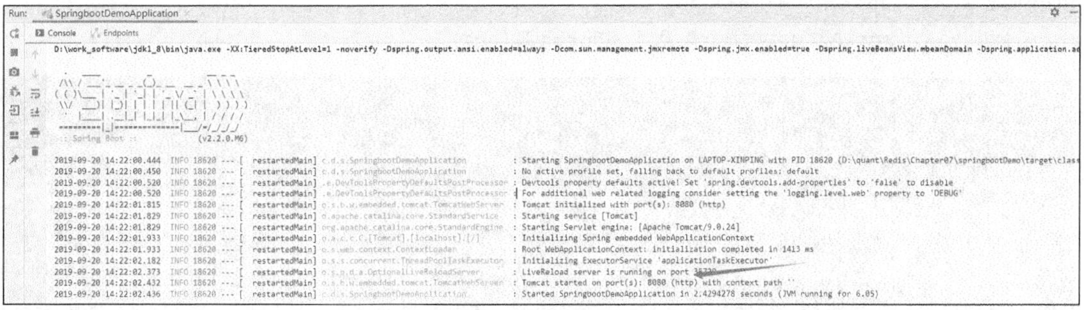

图 8-6 IntelliJ IDEA 控制台输出 Sprint Boot 的启动信息

如果要把项目部署到服务器上，还需要对项目进行发布。通过 Maven 的"mvn package"命令对整个 Spring Boot 项目进行编译、单元测试和打包，然后运行打包后的 JAR 文件。我们切换到项目所在的目录，然后执行"mvn package"命令，如图 8-7 所示。

图 8-7 编译 Spring Boot 项目

打包成功后，在工程的 target 目录中会生成 JAR 文件 springbootDemo-0.0.1-SNAPSHOT.jar，如图 8-8 所示。

图 8-8 打包成功后的 target 目录

在命令提示符窗口切换到 target 目录中，执行如下命令就可以启动项目，这样项目就发布成功了。启动的效果如图 8-9 所示。

```
java -jar springbootDemo-0.0.1-SNAPSHOT.jar
```

图 8-9　启动的效果

8.1.3　Spring Boot 结合 Redis 实战

8.1.2 小节讲了 Spring Boot 项目搭建与 Redis 整合应用，本小节讲解 Spring Boot 结合 Redis 实战。本实例目录为 Redis\Chapter08\springbootDemo，内容如下。

首先，建立 Maven 项目，在 Maven 项目中引入 pom.xml 文件。

```xml
<dependency>
    <groupId>org.springframework.boot</groupId>
    <artifactId>spring-boot-starter-data-redis</artifactId>
</dependency>
```

Jedis 封装了 Spring Boot 与 Redis 的连接工具。

其次，编写 RedisTemplate 对象，设置 RedisConnectionFactory 参数。本实例使用 "RedisConfig.java"，内容如下。

```java
package com.dxtd.springbootDemo;

import org.springframework.context.annotation.Bean;
import org.springframework.context.annotation.Configuration;
import org.springframework.data.redis.connection.RedisConnectionFactory;
import org.springframework.data.redis.core.RedisTemplate;

@Configuration
public class RedisConfig {
    @Bean
    public RedisTemplate<String,String> redisTemplate(RedisConnectionFactory factory){
        RedisTemplate<String,String> redisTemplate = new RedisTemplate<String,String>();
        redisTemplate.setConnectionFactory(factory);
        return redisTemplate;
    }

}
```

@Configuration 注解指明当前类是一个配置类，就是来代替当前的 Spring 配置文件。

编写控制器类 RedisController。本实例使用 "RedisController.java"，内容如下。

```java
package com.dxtd.springbootDemo.controller;

import org.springframework.data.redis.core.RedisTemplate;
import org.springframework.web.bind.annotation.RequestMapping;
import org.springframework.web.bind.annotation.ResponseBody;
import org.springframework.web.bind.annotation.RestController;
import javax.annotation.Resource;

@RestController
public class RedisController {
    @Resource
    private RedisTemplate redisTemplate;

    @RequestMapping("/redis/setAndGet")
    @ResponseBody
    public String setAndGetValue(String name,String value){
        redisTemplate.opsForValue().set(name,value);
        return (String) redisTemplate.opsForValue().get(name);
    }
}
```

@Resource 注解默认按照名称进行装配，名称可以通过 name 属性进行指定。如果没有指定 name 属性，则当注解写在字段上时，默认取字段名按照名称进行查找。在 RedisController 类中引入一个 Bean 命名为 RedisTemplate。

再次，引入配置文件 application.properties，在 springbootDemo\src\main\resources 下新建配置文件 application.properties，内容如下。

```
# Redis 数据库索引（默认为 0）
spring.redis.database=0

# Redis 服务器地址
spring.redis.host=127.0.0.1

# Redis 服务器连接端口
spring.redis.port=6379

# Redis 服务器连接密码（默认为空）
spring.redis.password=
```

最后，启动 SpringbootDemoApplication.java，访问控制器类 RedisController 对外提供的服务，在浏览器中访问 http://127.0.0.1:8080/redis/setAndGet?name=username&value=xinping，访问这个请求地址后，username 将作为 key，xinping 作为 value 保存到 Redis 中，并把这个 key-value 作为结果返回到前台页面，返回的结果是 JSON 格式的数据，如下所示。

```
{
    "username": "xinping"
}
```

在浏览器中访问这个请求的结果如图 8-10 所示。

图 8-10　返回消息

从以上结果可以看出，username 作为 key，xinping 作为 value，已经以 key-value 键值对的形式存入 Redis 了。

8.2　RedisTemplate API 详解

Redis 有 5 种不同的数据类型，这 5 种数据类型分别为 String、List、Set、Hash 和 Sorted Set。Spring 封装了 RedisTemplate 对象来对 5 种数据类型操作，它支持所有的 Redis 原生的 API。

- redisTemplate.opsForValue()：操作 String。
- redisTemplate.opsForHash()：操作 Hash。
- redisTemplate.opsForList()：操作 List。
- redisTemplate.opsForSet()：操作 Set。
- redisTemplate.opsForZSet()：操作 Sorted Set。

编写 RedisService 类，通过注解的方式调用 RedisTemplate 对象来操作 Redis 的 5 种数据类型。在 RedisService 类中使用了 @Service 注解来标注业务层，RedisService 类将自动注入到 Spring 容器中。本实例使用 "RedisService.java"，内容如下：

```
package com.dxtd.springbootDemo;

import org.springframework.beans.factory.annotation.Autowired;
import org.springframework.data.redis.core.*;
import org.springframework.stereotype.Service;

import java.io.Serializable;
import java.util.List;
import java.util.Set;
import java.util.concurrent.TimeUnit;

@Service
public class RedisService {
    @Autowired
    private RedisTemplate redisTemplate;

}
```

然后就可以在 RedisService 类里添加具体操作 Redis 的业务逻辑了。本节将详细介绍对 Redis 的 5 种数据类型的添加和获取操作进行封装。

8.2.1　写入和读取缓存

在 RedisService 类中封装写入 Redis 缓存的业务逻辑的具体内容如下，使用 redisTemplate.opsForValue() 操作 String。

```
public boolean set(final String key, Object value) {
    boolean result = false;
    try {
        ValueOperations<Serializable, Object> operations = redisTemplate.opsForValue();
        operations.set(key, value);
        result = true;
```

```java
    } catch (Exception e) {
        e.printStackTrace();
    }
    return result;
}
```

其中封装写入 Redis 缓存设置过期时间的业务逻辑的具体内容如下，需要调用 redisTemplate.expire(key, expireTime, TimeUnit.SECONDS)。

```java
public boolean set(final String key, Object value, Long expireTime) {
    boolean result = false;
    try {
        ValueOperations<Serializable, Object> operations = redisTemplate.opsForValue();
        operations.set(key, value);
        redisTemplate.expire(key, expireTime, TimeUnit.SECONDS);
        result = true;
    } catch (Exception e) {
        e.printStackTrace();
    }
    return result;
}
```

其中封装批量删除对应的 key 的业务逻辑的具体内容如下。

```java
public void remove(final String... keys) {
    for (String key : keys) {
        remove(key);
    }
}
```

其中封装删除对应的 key 的业务逻辑的具体内容如下，需要调用 redisTemplate.delete(key)。

```java
public void remove(final String key) {
    if (exists(key)) {
        redisTemplate.delete(key);
    }
}
```

其中封装判断 Redis 缓存中是否有对应的 key 的业务逻辑的具体内容如下，需要调用 redisTemplate.hasKey(key)。

```java
public boolean exists(final String key) {
    return redisTemplate.hasKey(key);
}
```

其中封装读取 Redis 缓存的业务逻辑的具体内容如下，使用 redisTemplate.opsForValue() 获得 operations 句柄，然后使用 operations.get(key) 获得 Redis 中对应的 key 的业务逻辑。

```java
public Object get(final String key) {
    Object result = null;
    ValueOperations<Serializable, Object> operations = redisTemplate.opsForValue();
    result = operations.get(key);
    return result;
}
```

8.2.2 添加和获取散列数据

在 RedisService 类中封装添加散列数据的业务逻辑的具体内容如下，使用 redisTemplate.

opsForHash()操作 Hash。

```java
public void hmSet(String key, Object hashKey, Object value) {
    HashOperations<String, Object, Object> hash = redisTemplate.opsForHash();
    hash.put(key, hashKey, value);
}
```

封装获取散列数据的业务逻辑的具体内容如下,使用 redisTemplate.opsForHash()获得 hash 句柄,然后使用 hash.get(key, hashKey)获得 key 对应的 Hash 值。

```java
public Object hmGet(String key, Object hashKey) {
    HashOperations<String, Object, Object> hash = redisTemplate.opsForHash();
    return hash.get(key, hashKey);
}
```

8.2.3 添加和获取列表数据

在 RedisService 类中封装添加列表数据的业务逻辑的具体内容如下,使用 redisTemplate.opsForList()操作 List。

```java
public void lPush(String k, Object v) {
    ListOperations<String, Object> list = redisTemplate.opsForList();
    list.rightPush(k, v);
}
```

封装获取列表数据的业务逻辑的具体内容如下,使用 redisTemplate.opsForList()获得 list 句柄,然后使用 list.range(k, l, l1)获得 Redis 的列表数据。

```java
public List<Object> lRange(String k, long l, long l1) {
    ListOperations<String, Object> list = redisTemplate.opsForList();
    return list.range(k, l, l1);
}
```

8.2.4 添加和获取集合数据

在 RedisService 类中封装添加集合数据的业务逻辑的具体内容如下,使用 redisTemplate.opsForSet() 操作 Set。

```java
public void add(String key, Object value) {
    SetOperations<String, Object> set = redisTemplate.opsForSet();
    set.add(key, value);
}
```

封装获取集合数据的业务逻辑的具体内容如下,使用 redisTemplate.opsForSet()获得 SetOperations<String, Object>对象,然后使用 set.members(key)获得 Redis 的集合数据。

```java
public Set<Object> setMembers(String key) {
    SetOperations<String, Object> set = redisTemplate.opsForSet();
    return set.members(key);
}
```

8.2.5 添加和获取有序集合数据

在 RedisService 类中封装添加有序集合数据的业务逻辑的具体内容如下,使用 redisTemplate.opsForZSet()操作 Sorted Set。

```java
public void zAdd(String key, Object value, double scoure) {
    ZSetOperations<String, Object> zset = redisTemplate.opsForZSet();
```

```
        zset.add(key, value, scoure);
    }
```
封装获取有序集合数据的业务逻辑的具体内容如下，使用 redisTemplate.opsForZSet()获得 ZSetOperations<String, Object>对象，然后使用 zset.rangeByScore(key, scoure, scoure1)获得 Redis 的有序集合数据。

```
    public Set<Object> rangeByScore(String key, double scoure, double scoure1) {
        ZSetOperations<String, Object> zset = redisTemplate.opsForZSet();
        redisTemplate.opsForValue();
        return zset.rangeByScore(key, scoure, scoure1);
    }
```

8.2.6　优化控制器

封装好 RedisService 类后就可以对控制器类 RedisController 进行优化了。

```
package com.dxtd.springbootDemo.controller;

import com.dxtd.springbootDemo.RedisService;
import org.springframework.data.redis.core.RedisTemplate;
import org.springframework.web.bind.annotation.RequestMapping;
import org.springframework.web.bind.annotation.ResponseBody;
import org.springframework.web.bind.annotation.RestController;
import javax.annotation.Resource;
import java.util.HashMap;
import java.util.Map;

@RestController
public class RedisController {
    @Resource
    private RedisTemplate redisTemplate;

    @Resource
    private RedisService service;

    @RequestMapping("/redis/setAndGet")
    @ResponseBody
    public Map setAndGetValue(String name, String value){
        System.out.println( "name="+ name + ",value=" + value );
        redisTemplate.opsForValue().set(name,value);
        Map result = new HashMap();
        String getValue =  (String) redisTemplate.opsForValue().get(name);
        result.put(name ,getValue);
        return result;
    }

    @RequestMapping("/redis/setAndGet2")
    @ResponseBody
    public Map setAndGetValueV2(String name,String value){
        service.set(name,value);
        Map result = new HashMap();
        String getValue =  service.get(name).toString();
```

```
        result.put(name ,getValue);
        return result;
    }
}
```

启动 SpringbootDemoApplication.java，使用 service.get()获得 Redis 缓存中的数据，访问 http://127.0.0.1:8080/redis/setAndGet2?name=db&value=redis，返回消息如图 8-11 所示。

图 8-11　返回消息

本节使用 RedisService 类使用 RedisTemplate API 操作 5 种数据类型，对 5 种数据类型的操作进行了封装。

8.3　Spring Boot 集成 Spring Session

传统 Session 的问题在于 Session 是由 Web 容器管理的，即一个 Session 只保存在一台服务器上，适合于单体应用。随着架构不断地向微服务分布式集群演进，传统的 Session 在集群环境下就不能正常工作了。例如，现在有 3 台 Web 服务器，客户端访问服务器通过负载均衡 Nginx 负载到某一台服务器上，用户此次的数据就保存到这台服务器的 Web 容器中了，如果用户下次请求被负载到其他服务器上，就获取不到之前保存的数据了。

这时候就需要整个服务器集群共享同一个 Session。为了解决所有服务器共享一个 Session 的问题，Session 就不能单独保存在自己的 Web 容器中，而是保存在一个公共的会话仓库（Session Repository）中，所有服务器都访问同一个会话仓库，这样所有服务器的状态都一致了。Spring Session 支持的会话仓库有 Redis、MongoDB、JDBC，本章使用 Redis 作为 Spring Session 的会话仓库。

Spring Session 有以下优点。

- Spring Session 是基于 Servlet 规范实现的一套 Session 管理框架。Spring Session 主要解决了分布式场景下 Session 的共享问题。Spring Session 最核心的类是 SessionRepositoryFilter，用于包装用户的请求和响应。
- 可在程序中直接替换 HttpSession，而无须修改一行代码。
- 可以很方便地与 Spring Security 集成，增加诸如 findSessionsByUserName、rememberMe，限制同一个账号可以同时在线的 Session 数（例如设置成 1，即可达到把前一次登录替换的效果）等。

8.3.1　配置 Spring Boot 项目

在 https://start.spring.io/中创建一个 Spring Boot 项目 SpringSessionDemo，使用 IntelliJ IDEA 导入项目。本实例目录为 Redis\Chapter08\SpringSessionDemo，内容如下。

在 Maven 项目的 pom.xml 文件里引入必要的依赖包。

```xml
<dependency>
    <groupId>org.springframework.boot</groupId>
    <artifactId>spring-boot-starter-web</artifactId>
</dependency>

<dependency>
    <groupId>org.springframework.boot</groupId>
    <artifactId>spring-boot-starter-web</artifactId>
</dependency>

<dependency>
    <groupId>org.springframework.session</groupId>
    <artifactId>spring-session-data-redis</artifactId>
</dependency>

<dependency>
    <groupId>org.springframework.boot</groupId>
    <artifactId>spring-boot-starter-data-redis</artifactId>
</dependency>
```

新建 Spring Boot 的配置文件 application.properties，内容如下。

```
# Redis 配置
# Redis 数据库索引（默认为 0）
spring.redis.database=0
# Redis 服务器地址
spring.redis.host=127.0.0.1
# Redis 服务器连接端口
spring.redis.port=6379
# Redis 服务器连接密码（默认为空）
spring.redis.password=
```

8.3.2 创建配置类和控制器类

创建配置类 RedisHttpSessionConfiguration，本实例使用"Configuration.java"，内容如下。

```java
package com.dxtd.SpringSessionDemo;

import org.springframework.context.annotation.Configuration;
import org.springframework.session.data.redis.config.annotation.web.http.EnableRedisHttpSession;

@Configuration
//maxInactiveIntervalInSeconds 默认是 1800s 过期，这里测试修改为 60s
@EnableRedisHttpSession(maxInactiveIntervalInSeconds=60)
public class RedisHttpSessionConfiguration {

}
```

配置类 RedisHttpSessionConfiguration 使用@Configuration 注解表明这是一个配置类。在这个类中也添加了注解@EnableRedisHttpSession，表示开启 Redis 的 Session 管理。如果需要设置会话失效时间，可以使用@EnableRedisHttpSession(maxInactiveIntervalInSeconds = 60)表示在 60 秒后会话失效。

创建用户类 User，用户类只有两个属性 username 和 password，保存用户登录的用户名和密码。本实例使用"User.java"，内容如下。

```java
package com.dxtd.SpringSessionDemo.controller;

import java.io.Serializable;

public class User implements Serializable{
    private String name;
    private String password;

    public String getName() {
        return name;
    }

    public void setName(String name) {
        this.name = name;
    }

    public String getPassword() {
        return password;
    }

    public void setPassword(String password) {
        this.password = password;
    }

    @Override
    public String toString() {
        return "User{" +
                "name='" + name + '\'' +
                ", password='" + password + '\'' +
                '}';
    }
}
```

创建控制器类 SessionController，在这个类中定义了 2 个方法。login 方法用于登录验证，当输入 username 等于"xinping"，password 等于"123"，判断用户登录成功，在 Session 中保存用户信息。get 方法用于从 Session 中获取用户信息。本实例使用"SessionController.java"，内容如下。

```java
package com.dxtd.SpringSessionDemo.controller;

import org.springframework.web.bind.annotation.*;
import javax.servlet.http.HttpServletRequest;
import javax.servlet.http.HttpSession;
import java.util.HashMap;
import java.util.Map;

@RestController
public class SessionController {
```

```java
/**
 * 登录系统
 *
 * */
@RequestMapping(value="/session/login", method = RequestMethod.GET)
@ResponseBody
public Map login(@RequestParam("username") String username,@RequestParam("password") String password,HttpServletRequest request, HttpSession session) {
    String message = "login failure";
    Map<String, Object> result = new HashMap<String, Object>();
    if(username != null && "xinping".equals(username) && "123".equals(password)){
        User user = new User();
        user.setName(username);
        user.setPassword(password);
        request.getSession().setAttribute("admin", user);
        message = "login success";
        result.put("message" , message);
        result.put("sessionId",session.getId());
    }else{
        result.put("message" , message);
    }

    return result;
}

/**
 * 查询用户
 *
 * */
@RequestMapping(value="/session/get", method = RequestMethod.GET)
@ResponseBody
public Map get(@RequestParam("username") String username,HttpServletRequest request, HttpSession session) {
    Object value = request.getSession().getAttribute("admin");
    Map<String,Object> result = new HashMap<String,Object>();
    result.put("message" ,value);
    result.put("sessionId",session.getId());
    return result;
}

}
```

8.3.3 编译和部署项目

启动项目，运行 SpringSessionDemoApplication.java，进入 SpringSessionDemo 项目所在目录。

```
D:\>cd D:\quant\Redis\Chapter08\SpringSessionDemo
```

使用 Maven 的 "mvn clean package" 命令将 SpringSessionDemo 项目打包成 JAR 文件。

```
D:\quant\Redis\Chapter08\SpringSessionDemo>mvn clean package
```

在 SpringSessionDemo\target 目录下获得编译好的压缩包 SpringSessionDemo-0.0.1-SNAPSHOT.jar，如图 8-12 所示。

图 8-12　编译好的压缩包

使用如下命令分别以两个不同的端口（8081 和 8082）启动项目，用于模拟分布式应用中的两个服务。

```
java -jar SpringSessionDemo-0.0.1-SNAPSHOT.jar --server.port=8081
java -jar SpringSessionDemo-0.0.1-SNAPSHOT.jar --server.port=8082
```

启动项目，如图 8-13 所示。

图 8-13　分别以两个不同的端口启动项目

第 1 步：访问 http://127.0.0.1:8081/session/login?username=xinping&password=123。

第一次访问 URL 地址端口为 8081 的 Spring Boot 项目时模拟用户登录，在浏览器中访问这个请求后的结果如图 8-14 所示。

图 8-14　第一次访问请求后的结果

查看 Redis 中的 key 可以看出有一个 key 为"spring:session:sessions:ddf10226-8ffe-4294-ab93-7a461535f8ed"，这个 key 包含了图 8-14 的页面中显示的 Session ID 值，说明 Session 已经成功保存到了 Redis 中，如图 8-15 所示。

图 8-15　查询 Redis 中的 key

第 2 步：访问地址 http://127.0.0.1:8082/session/get?username=xinping。

第二次访问 URL 地址端口为 8082 的 Spring Boot 项目时模拟获取登录的用户信息，在浏览器中访问这个请求后的结果如图 8-16 所示。

图 8-16　第二次访问请求的结果

可以看出成功获取了保存在 Redis 中的用户名为 xinping 的用户信息，并且两次请求的 Session ID 是相同的，实现了 Session 的共享。

通过 Spring Boot 使用 Redis 来实现 Session 的共享很方便，再配合 Nginx 进行负载均衡，便能实现分布式的应用了。

第 9 章 Redis 监控

Redis 作为 Web 应用开发的黄金搭档被大量应用，广泛地用于存储 Session 信息、交易数据、社交网络等热数据。为了保证系统的稳定性，业务需要 Redis 状态监控运维。我们可以使用 Redis 自带的 info 命令和开源的 Redis 监控工具，如果还不能满足业务需求，我们可以开发 Redis 自定义监控系统。

9.1 Redis 监控指标

9.1.1 使用 INFO 命令

INFO 命令会返回 Redis 服务器的状态信息和统计数据，使用 INFO 命令可以监控 Redis 的工作状态。可以通过以下可选参数（可选参数不区分大小写），选择查看特定分段的 Redis 服务器信息，比如 INFO server，INFO clients 等。

- server：Redis 服务器相关的通用信息。
- clients：客户端连接的相关信息。
- memory：内存消耗的相关信息。
- persistence：RDB（Redis Database）和 AOF（Append-Only File）的相关信息。
- stats：通用统计数据。
- replication：主/从复制的相关信息。
- cpu：CPU 消耗的统计数据。
- cluster：Redis 集群的相关信息。
- keyspace：数据库相关的统计数据。

进入 Redis 客户端输入 INFO 命令，会返回批量字符串，是多行文本行的集合。文本行包含一个分段名称，以一个#字符开始下面显示属性，所有的属性都是以 field:value 的格式显示，以\r\n 结尾。INFO 命令的返回值如下所示。

```
127.0.0.1:6379> INFO
# Server
redis_version:6.0.6
redis_git_sha1:00000000
redis_git_dirty:0
redis_build_id:1c5fdf3388fb1cbf
```

```
redis_mode:standalone
os:Linux 4.14.231-173.361.amzn2.x86_64 x86_64
arch_bits:64
multiplexing_api:epoll
atomicvar_api:atomic-builtin
gcc_version:7.3.1
process_id:6970
run_id:19beba68248e699d5ca388d67913ba5795da43da
tcp_port:6379
uptime_in_seconds:15088
uptime_in_days:0
hz:10
configured_hz:10
lru_clock:11766283
executable:/soft/redis-6.0.6/redis-server
config_file:/usr/local/redis/conf/redis.conf

# Clients
connected_clients:1
client_recent_max_input_buffer:2
client_recent_max_output_buffer:0
blocked_clients:0
tracking_clients:0
clients_in_timeout_table:0

# Memory
used_memory:866648
used_memory_human:846.34K
used_memory_rss:3489792
used_memory_rss_human:3.33M
used_memory_peak:866680
used_memory_peak_human:846.37K
used_memory_peak_perc:100.00%
used_memory_overhead:820146
used_memory_startup:803016
used_memory_dataset:46502
used_memory_dataset_perc:73.08%
allocator_allocated:1119000
allocator_active:1478656
allocator_resident:4292608
total_system_memory:32891297792
total_system_memory_human:30.63G
used_memory_lua:37888
used_memory_lua_human:37.00K
used_memory_scripts:0
used_memory_scripts_human:0B
number_of_cached_scripts:0
maxmemory:0
maxmemory_human:0B
maxmemory_policy:noeviction
```

```
allocator_frag_ratio:1.32
allocator_frag_bytes:359656
allocator_rss_ratio:2.90
allocator_rss_bytes:2813952
rss_overhead_ratio:0.81
rss_overhead_bytes:-802816
mem_fragmentation_ratio:4.23
mem_fragmentation_bytes:2665656
mem_not_counted_for_evict:0
mem_replication_backlog:0
mem_clients_slaves:0
mem_clients_normal:16986
mem_aof_buffer:0
mem_allocator:jemalloc-5.1.0
active_defrag_running:0
lazyfree_pending_objects:0

# Persistence
loading:0
rdb_changes_since_last_save:0
rdb_bgsave_in_progress:0
rdb_last_save_time:1622363931
rdb_last_bgsave_status:ok
rdb_last_bgsave_time_sec:-1
rdb_current_bgsave_time_sec:-1
rdb_last_cow_size:0
aof_enabled:0
aof_rewrite_in_progress:0
aof_rewrite_scheduled:0
aof_last_rewrite_time_sec:-1
aof_current_rewrite_time_sec:-1
aof_last_bgrewrite_status:ok
aof_last_write_status:ok
aof_last_cow_size:0
module_fork_in_progress:0
module_fork_last_cow_size:0

# Stats
total_connections_received:2
total_commands_processed:24
instantaneous_ops_per_sec:0
total_net_input_bytes:1488
total_net_output_bytes:55040
instantaneous_input_kbps:0.00
instantaneous_output_kbps:0.00
rejected_connections:0
sync_full:0
sync_partial_ok:0
sync_partial_err:0
expired_keys:0
expired_stale_perc:0.00
```

```
expired_time_cap_reached_count:0
expire_cycle_cpu_milliseconds:174
evicted_keys:0
keyspace_hits:0
keyspace_misses:0
pubsub_channels:0
pubsub_patterns:0
latest_fork_usec:0
migrate_cached_sockets:0
slave_expires_tracked_keys:0
active_defrag_hits:0
active_defrag_misses:0
active_defrag_key_hits:0
active_defrag_key_misses:0
tracking_total_keys:0
tracking_total_items:0
tracking_total_prefixes:0
unexpected_error_replies:0

# Replication
role:master
connected_slaves:0
master_replid:32e87cdc7ba8f2a6eedfb10a1be8df27813a1a99
master_replid2:0000000000000000000000000000000000000000
master_repl_offset:0
second_repl_offset:-1
repl_backlog_active:0
repl_backlog_size:1048576
repl_backlog_first_byte_offset:0
repl_backlog_histlen:0

# CPU
used_cpu_sys:4.558479
used_cpu_user:7.538676
used_cpu_sys_children:0.000000
used_cpu_user_children:0.000000

# Cluster
cluster_enabled:0

# Keyspace
db0:keys=2,expires=0,avg_ttl=0
```

1. server 分段的字段含义

```
127.0.0.1:6379> INFO server
# Server
redis_version:6.0.6
redis_git_sha1:00000000
redis_git_dirty:0
redis_build_id:1c5fdf3388fb1cbf
```

```
redis_mode:standalone
os:Linux 4.14.231-173.361.amzn2.x86_64 x86_64
arch_bits:64
multiplexing_api:epoll
atomicvar_api:atomic-builtin
gcc_version:7.3.1
process_id:30633
run_id:5b93f95d05073508b29296e2ca25263bb52c90e9
tcp_port:6379
uptime_in_seconds:546
uptime_in_days:0
hz:10
configured_hz:10
lru_clock:11661872
executable:/soft/redis-6.0.6/redis-server
config_file:/usr/local/redis/conf/redis.conf
```

server 分段包含关于 Redis 服务器相关的信息,server 分段中的主要字段的含义如下所示。

- redis_version:Redis 服务器版本。
- redis_git_sha1:Git SHA1。
- redis_git_dirty:Git dirty flag。
- redis_mode:运行模式(单机或集群)。
- os:Redis 服务器的宿主操作系统。
- arch_bits:架构(32 或 64 位)。
- multiplexing_api:Redis 所使用的事件处理机制。
- atomicvar_api:Redis 使用的 Atomicvar API。
- gcc_version:用来编译这个 Redis 服务器的 GCC 编译器的版本。
- process_id:Redis 服务器进程的进程号(PID)。
- run_id:用于标识 Redis 服务器的随机标识符(Redis 的哨兵模式和集群模式会使用这个随机值)。
- tcp_port:TCP/IP 的监听端口。
- uptime_in_seconds:自 Redis 服务器启动以来,经过的秒数。
- uptime_in_days:自 Redis 服务器启动以来,经过的天数。
- lru_clock:以分钟为单位进行自增的时钟,用于 LRU(Least Recently Used,最近最少使用)缓存管理管理。
- executable:Redis 服务器的可执行文件的路径。
- config_file:Redis 服务器读取的配置文件的路径。

2. clients 分段的字段含义

```
127.0.0.1:6379> INFO clients
# Clients
connected_clients:1
client_recent_max_input_buffer:2
client_recent_max_output_buffer:0
blocked_clients:0
```

```
tracking_clients:0
clients_in_timeout_table:0
```
clients 分段包含已连接的 Redis 客户端信息，clients 分段中的所有字段的含义如下所示。

- connected_clients：已连接客户端的数量。
- client_recent_max_input_buffer：当前连接的客户端当中，最大输入缓存。
- client_recent_max_output_buffer：当前连接的客户端当中，最大输出缓存。
- blocked_clients：正在等待阻塞命令（BLPOP、BRPOP、BRPOPLPUSH）的客户端的数量。
- tracking_clients：被跟踪的客户数量。
- clients_in_timeout_table：客户端超时表中的客户端数量。

3. memory 分段的字段含义

```
127.0.0.1:6379> INFO memory
# Memory
used_memory:865128
used_memory_human:844.85K
used_memory_rss:2813952
used_memory_rss_human:2.68M
used_memory_peak:866616
used_memory_peak_human:846.30K
used_memory_peak_perc:99.83%
used_memory_overhead:820114
used_memory_startup:803016
used_memory_dataset:45014
used_memory_dataset_perc:72.47%
allocator_allocated:970680
allocator_active:1265664
allocator_resident:4079616
total_system_memory:32891297792
total_system_memory_human:30.63G
used_memory_lua:37888
used_memory_lua_human:37.00K
used_memory_scripts:0
used_memory_scripts_human:0B
number_of_cached_scripts:0
maxmemory:0
maxmemory_human:0B
maxmemory_policy:noeviction
allocator_frag_ratio:1.30
allocator_frag_bytes:294984
allocator_rss_ratio:3.22
allocator_rss_bytes:2813952
rss_overhead_ratio:0.69
rss_overhead_bytes:-1265664
mem_fragmentation_ratio:3.41
mem_fragmentation_bytes:1989840
mem_not_counted_for_evict:0
mem_replication_backlog:0
```

```
mem_clients_slaves:0
mem_clients_normal:16986
mem_aof_buffer:0
mem_allocator:jemalloc-5.1.0
active_defrag_running:0
lazyfree_pending_objects:0
```

memory 分段包含 Redis 服务器的内存信息，memory 分段中的主要字段的含义如下所示。

- used_memory：Redis 分配的内存总量，以字节（Byte）为单位。
- used_memory_human：以用户可读的格式返回 Redis 分配的内存总量，以千字节（KB）为单位。
- used_memory_rss：从操作系统的角度，返回 Redis 已分配的内存总量（也被称为"常驻集大小"）。这个值和 top、ps 等命令的输出一致，以 Byte 为单位。
- used_memory_rss_human：以用户可读的格式返回 Redis 已分配的内存总量，以兆字节（MB）为单位。
- used_memory_peak：Redis 消耗的的内存峰值，以 Byte 为单位。
- used_memory_peak_human：以用户可读的格式返回 Redis 消耗的内存峰值，以 KB 为单位。
- used_memory_peak_perc：used_memory_peak 占 used_memory 的百分比。
- used_memory_overhead：Redis 服务器分配用于管理其内部数据结构的所有开销的总和，以 Byte 为单位。
- used_memory_startup：Redis 在启动时消耗的初始内存量，以 Byte 为单位。
- allocator_allocated：Redis 分配器分配的内存。
- allocator_active：Redis 分配器活跃的内存。
- allocator_resident：Redis 分配器常驻的内存。
- total_system_memory：Redis 主机拥有的总内存量，以 Byte 为单位。
- total_system_memory_human：以用户可读的格式返回 Redis 主机拥有的总内存量，以吉字节（GB）为单位。
- used_memory_lua：Lua 引擎所使用的内存大小，以 Byte 为单位。
- used_memory_lua_human：以用户可读的格式返回引擎所使用的内存大小，以 GB 为单位。
- used_memory_scripts：缓存 Lua 脚本使用的字节数。
- used_memory_scripts_human：以用户可读的格式返回缓存 Lua 脚本使用的字节数。
- maxmemory：Redis 能使用的最大内存上限，0 表示没有限制，以 Byte 为单位。
- maxmemory_human：以用户可读的格式返回最大内存上限。
- maxmemory_policy：Redis 使用的的内存回收策略，可以是 noeviction、allkeys-lru、volatile-lru、allkeys-random、volatile-random 或 volatile-ttl。
- mem_fragmentation_ratio：used_memory_rss 和 used_memory 之间的比率。
- mem_allocator：Redis 使用的内存分配器。
- active_defrag_running：指示活动碎片整理是否处于活动状态的标志。
- lazyfree_pending_objects：等待被释放的对象数（由于调用 UNLINK 或带有 ASYNC

选项的 FLUSHDB 和 FLUSHALL）。

在理想情况下，used_memory_rss 的值应该只比 used_memory 稍微高一点儿。当常驻集内存（rss）远大于已使用内存（used），这表示 Redis 服务器存在（内部或外部的）内存碎片。内存碎片的比率可以通过 mem_fragmentation_ratio 的值看出。当已使用内存（used）远大于常驻集内存（rss），这表示 Redis 的一部分内存被操作系统换出到磁盘了，在这种情况下，操作可能会产生明显的延迟。

当 Redis 释放内存时，分配器可能会，也可能不会，将内存返还给操作系统。如果 Redis 释放了内存，却没有将内存返还给操作系统，那么 used_memory 的值可能和操作系统显示的 Redis 内存占用并不一致。这可能是由于 Redis 正在使用和释放内存，但是释放的内存尚未归还给操作系统而导致的。查看 used_memory_peak 的值可以验证这种情况是否发生。

4. persistence 分段的字段含义

```
127.0.0.1:6379> INFO persistence
# Persistence
loading:0
rdb_changes_since_last_save:0
rdb_bgsave_in_progress:0
rdb_last_save_time:1622274963
rdb_last_bgsave_status:ok
rdb_last_bgsave_time_sec:0
rdb_current_bgsave_time_sec:-1
rdb_last_cow_size:163840
aof_enabled:0
aof_rewrite_in_progress:0
aof_rewrite_scheduled:0
aof_last_rewrite_time_sec:-1
aof_current_rewrite_time_sec:-1
aof_last_bgrewrite_status:ok
aof_last_write_status:ok
aof_last_cow_size:0
module_fork_in_progress:0
module_fork_last_cow_size:0
```

persistence 分段包含 RDB 和 AOF 的相关信息，persistence 分段中的所有字段的含义如下所示。

- loading：表示 Redis 是否正在加载一个转储文件的标志。
- rdb_changes_since_last_save：从最近一次转储至今，RDB 的修改次数。
- rdb_bgsave_in_progress：表示 Redis 正在保存 RDB 的标志。
- rdb_last_save_time：最近一次成功保存 RDB 的时间戳，基于 Epoch 时间。
- rdb_last_bgsave_status：最近一次 RDB 保存操作的状态。
- rdb_last_bgsave_time_sec：最近一次 RDB 保存操作的持续时间，以秒（s）为单位。
- rdb_current_bgsave_time_sec：如果 Redis 正在执行 RDB 保存操作，那么这个字段表示已经消耗的时间，以 s 为单位。
- rdb_last_cow_size：最近一次 RDB 保存操作期间写时复制内存的大小，以 Byte 为单位。

- aof_enabled：表示 AOF 日志记录已激活的标志。
- aof_rewrite_in_progress：表示 AOF 重写操作正在进行的标志。
- aof_rewrite_scheduled：表示一旦 Redis 正在执行的 RDB 保存操作完成之后，是否就会调度执行 AOF 重写操作的标志。
- aof_last_rewrite_time_sec：最近一次 AOF 重写操作的持续时间，以 s 为单位。
- aof_current_rewrite_time_sec：如果 Redis 正在执行 AOF 重写操作，那么这个字段表示已经消耗的时间，以 s 为单位。
- aof_last_bgrewrite_status：最近一次 AOF 重写操作的状态。
- aof_last_write_status：对 AOF 的最后一次写操作的状态。
- aof_last_cow_size：最近一次 AOF 重写操作期间写时复制内存的大小，以 Byte 为单位。
- module_fork_in_progress：表示模块分叉正在进行的标志。
- module_fork_last_cow_size：最近一次模块 fork 操作期间写时复制内存的大小，以 Byte 为单位。

5．stats 分段的字段含义

```
127.0.0.1:6379> INFO stats
# Stats
total_connections_received:1
total_commands_processed:4
instantaneous_ops_per_sec:0
total_net_input_bytes:117
total_net_output_bytes:3594
instantaneous_input_kbps:0.00
instantaneous_output_kbps:0.00
rejected_connections:0
sync_full:0
sync_partial_ok:0
sync_partial_err:0
expired_keys:0
expired_stale_perc:0.00
expired_time_cap_reached_count:0
expire_cycle_cpu_milliseconds:0
evicted_keys:0
keyspace_hits:0
keyspace_misses:0
pubsub_channels:0
pubsub_patterns:0
latest_fork_usec:0
migrate_cached_sockets:0
slave_expires_tracked_keys:0
active_defrag_hits:0
active_defrag_misses:0
active_defrag_key_hits:0
active_defrag_key_misses:0
tracking_total_keys:0
tracking_total_items:0
```

```
tracking_total_prefixes:0
unexpected_error_replies:0
```
stats 分段包含 Redis 的一般统计信息，stats 分段中的主要字段的含义如下所示。

- total_connections_received：Redis 服务器接收的连接总数。
- total_commands_processed：Redis 服务器处理的命令总数。
- instantaneous_ops_per_sec：Redis 服务器每秒处理的命令数。
- total_net_input_bytes：Redis 服务器从网络读取的总字节数。
- total_net_output_bytes：Redis 服务器写入网络的总字节数。
- instantaneous_input_kbps：每秒读取数据的速率，以 KB/s 为单位。
- instantaneous_output_kbps：每秒钟发送数据的速率，以 KB/s 为单位。
- rejected_connections：Redis 服务器由于 maxclients 限制而拒绝的连接数量。
- sync_full：Redis 主机和从机进行完全同步的次数。
- sync_partial_ok：Redis 服务器接受 PSYNC 请求的次数。
- sync_partial_err：Redis 服务器拒绝 PSYNC 请求的次数。
- expired_keys：键过期事件的总数。
- expired_stale_perc：由于 maxmemory 限制，而被回收内存的键的总数。
- evicted_keys：由于 maxmemory 限制，而被回收内存的键的总数。
- keyspace_hits：在主字典中成功查找键的次数。
- keyspace_misses：在主字典中键查找失败的次数。
- pubsub_channels：客户端订阅的发布/订阅频道的总数量。
- pubsub_patterns：客户端订阅的发布/订阅模式的总数量。
- latest_fork_usec：最近一次 fork 操作消耗的时间，以微秒（ms）为单位。
- migrate_cached_sockets：迁移已缓存的套接字的数量。

6. replication 分段的字段含义

```
127.0.0.1:6379> INFO replication
# Replication
role:master
connected_slaves:0
master_replid:32e87cdc7ba8f2a6eedfb10a1be8df27813a1a99
master_replid2:0000000000000000000000000000000000000000
master_repl_offset:0
second_repl_offset:-1
repl_backlog_active:0
repl_backlog_size:1048576
repl_backlog_first_byte_offset:0
repl_backlog_histlen:0
```
replication 分段包含 Redis 的主/从复制信息，replication 分段中的主要字段的含义如下所示。

- role：如果这个 Redis 实例是某个 Redis 主机的从机，那么这个字段的值为 "slave"。
如果这个实例不是任何 Redis 主机的从机，那么这个字段的值是 "master"。
- connected_slaves：已连接的 Redis 从机的数量。
- master_replid：Redis 服务器的复制 ID。

- master_replid2：辅助复制 ID，用于故障转移后的 PSYNC。
- master_repl_offset：Redis 服务器的当前复制偏移量。
- second_repl_offset：接受复制 ID 的偏移量。
- repl_backlog_active：表示 Redis 服务器是否为部分同步开启复制备份日志（backlog）功能的标志。
- repl_backlog_size：备份日志的循环缓冲区的大小。
- repl_backlog_first_byte_offset：备份日志缓冲区中的首个字节的复制偏移量。
- repl_backlog_histlen：备份日志的实际数据长度。

7. cpu 分段的字段含义

```
127.0.0.1:6379> INFO cpu
# CPU
used_cpu_sys:3.919009
used_cpu_user:6.572067
used_cpu_sys_children:0.000000
used_cpu_user_children:0.000000
```

cpu 分段包含 Redis 使用的 CPU 统计信息，cpu 分段中的所有字段的含义如下所示。

- used_cpu_sys：Redis 服务器消耗的系统 CPU 性能。
- used_cpu_user：Redis 服务器消耗的用户 CPU 性能。
- used_cpu_sys_children：后台进程消耗的系统 CPU 性能。
- used_cpu_user_children：后台进程消耗的用户 CPU 性能。

8. cluster 分段的字段含义

```
127.0.0.1:6379> INFO cluster
# Cluster
cluster_enabled:0
```

cluster 分段包含 Redis 集群信息，cluster 分段当前只包含一个字段。

cluster_enabled：表示是否启用 Redis 集群功能的标志。

9. keyspace 分段的字段含义

```
127.0.0.1:6379> INFO keyspace
# Keyspace
db0:keys=2,expires=0,avg_ttl=0
```

keyspace 分段会提供关于每个数据库的主字典的统计数据。这些统计数据包括键的数量、具有过期时间的键的数量和键的平均生存时间。

对于每个 Redis 数据库来说，keyspace 分段每行的格式如下所示。

```
dbXXX: keys=XXX,expires=XXX
```

其中，第一个 XXX 表示数据库的编号，第二个 XXX 表示键的数量，第三个 XXX 表示具有过期时间的键的数量，第四个 XXX 表示键的平均生存时间。

9.1.2 使用 redis-stat

redis-stat 是基于 Ruby 实现的一个简单易用的 Redis 监控工具。redis-stat 的实现原理基于

Redis 的 INFO 命令，相比 Redis 的 MONITOR 命令来说其对 Redis 不会产生任何性能影响。

redis-stat 的使用需要依赖 Ruby 环境，但 JRuby 为我们提供了基于 Java 的使用方式。在 GitHub 官网搜索 redis-stat，从 redis-stat 项目主页下载 redis-stat-0.4.14.jar 文件，如图 9-1 所示。

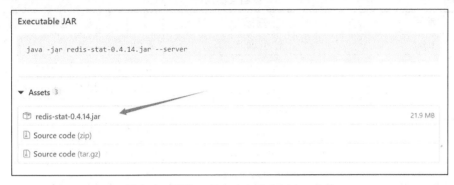

图 9-1　下载 redis-stat-0.4.14.jar 文件

执行启动脚本。

```
java -jar redis-stat-0.4.14.jar 127.0.0.1:6379 3 --server=63800 > monitor.log &
```

启动脚本表示通过 http://127.0.0.1:63800 访问 redis-stat，监控 IP 地址为 127.0.0.1，端口为 6379 的 Redis 服务器，并把日志输出到 monitor.log 文件中。

通过 Web 页面直接访问 http://127.0.0.1:63800，如图 9-2 所示。

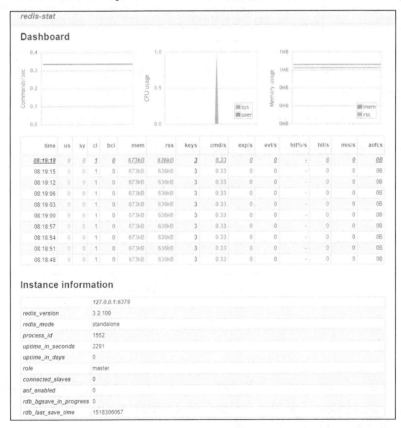

图 9-2　redis-stat 页面

页面中的图形是根据 Redis 服务器的状态动态变化的，还是可以满足一般的监控需求的。页面指标说明如表 9-1 所示。

表 9-1　　　　　　　　　　　　　页面指标说明

简写	指标	说明
time	time	查询时间
us	used_cpu_user	用户空间占用 CPU 的百分比
sy	used_cpu_sys	内核空间占用 CPU 的百分比
cl	connected_clients	连接客户端的数量
bcl	blocked_clients	阻塞客户端的数量（如 BLPOP）
mem	used_memory	使用总内存
rss	used_memory_rss	使用物理内存
keys	dbx.keys	key 的总数量
cmd/s	command/s	每秒执行命令数
exp/s	expired_keys/s	每秒过期 key 数量
evt/s	evicted_keys/s	每秒淘汰 key 数量
hit%/s	keyspace_hitratio/s	每秒命中百分比
hit/s	keyspace_hits/s	每秒命中数量
mis/s	keyspace_miss/s	每秒丢失数量
aofcs	aof_current_size	AOF 日志当前大小

按"F12"键打开浏览器 Chrome 的控制台，切换到网络选项，可以看到每 5～8s 刷新了一次页面，如图 9-3 所示。

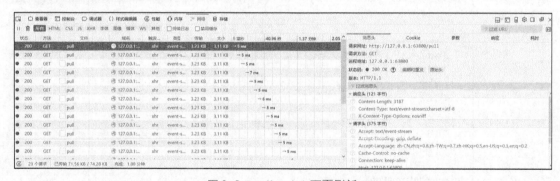

图 9-3　redis-stat 页面刷新

9.2　自定义监控

redis-stat 提供了一种简易的方式实现对 Redis 的监控，但多数场景下可能需要进行定制，例如鉴权方式、统计指标、告警等。为了实现更灵活的控制，可利用类似的方式对 INFO 信息进行解析，以实现对 Redis 的统一监控。

在本节我们使用 Spring Boot 做一个自定义 Redis 监控平台，前端使用 Bootstrap 3 页面展示 Redis 的监控指标，使用 WebSocket 技术实时监控 Redis 数据库 db0 的 keys 数量，在页面中使用曲线图形动态地显示 keys 数量。完整的自定义监控页面如图 9-4 所示。

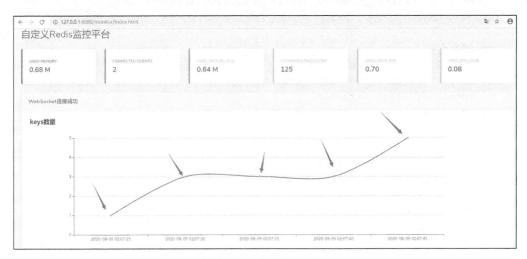

图 9-4　自定义监控页面

从图 9-4 中可以看出自定义监控页面中最多显示 5 个时间点的 key 数量。使用 IntelliJ IDEA 新建一个 Spring Boot 项目，命名为 RedisMonitor，本实例目录为 Redis\Chapter09\RedisMonitor，项目目录结构如图 9-5 所示。

图 9-5　自定义监控项目的目录结构

在监控项目 RedisMonitor 的 pom.xml 文件里引入必要的依赖包。

```xml
<dependency>
    <groupId>org.springframework.boot</groupId>
    <artifactId>spring-boot-starter-websocket</artifactId>
</dependency>

<dependency>
    <groupId>com.alibaba</groupId>
    <artifactId>fastjson</artifactId>
    <version>1.2.71</version>
</dependency>

<dependency>
    <groupId>redis.clients</groupId>
    <artifactId>jedis</artifactId>
    <version>3.1.0</version>
</dependency>
```

9.2.1 前端页面

1. Bootstrap 3 后台管理系统模板

Redis 自定义监控平台的前端页面使用的是开源的模板 Start Bootstrap Admin 2，它是基于 Bootstrap 3 开发的，是一个轻量级的后台管理系统模板。Start Bootstrap Admin 2 有强大的功能组件和 UI 组件，基本能满足后台管理系统的需求，我们在此基础上进行二次开发，如图 9-6 所示。读者可以从 Start Bootstrap Admin 2 的官网下载最新的源码包，把源码包压缩后放在监控项目 RedisMonitor 的\src\main\resources\static 目录下。

图 9-6　Start Bootstrap Admin 2 模板有强大的功能组件和 UI 组件

2. ECharts

Redis 自定义监控平台使用的图表是 ECharts。ECharts 是开源的商业及数据图表库，它是一个纯 JavaScript 的图表库，可以流畅地运行在 PC 和移动设备上，兼容当前绝大部分浏览器（IE 8/9/10/11、Chrome、Firefox、Safari 等），底层依赖轻量级的 Canvas 类库 ZRender，提供直观、生动、可交互、可高度定制化的数据可视化图表。ECharts 中加入了更丰富的交互功能以及更多的可视化效果，并对移动端做了深度的优化。

ECharts 是由百度 EFE 数据可视化团队开发的，是基于 JavaScript 的数据图表库，在编程的灵活性和图表的丰富性方面非常强大，优点很多。

- ECharts 是一款独立的 Web 版数据可视化工具，界面友好，提供强大的互动性操作。
- 对图形参数的修改十分简单、直观，便于初学者使用。
- 丰富的可视化图表，具有高度互动性，这得益于其完善的文档和简单的 JavaScript API。相比 Matplotlib 的图表，ECharts 更加现代和绚丽。
- 深度的交互式数据探索。ECharts 提供了图例、视觉映射、数据区域缩放、Tooltip、数据筛选等开箱即用的交互组件，可以对数据进行多维度筛取、视图缩放、展示细节等交互操作。
- 移动端优化。

ECharts 提供了常规的折线图、柱状图、散点图、饼图、K 线图、用于统计的盒形图、用于地理数据可视化的地图、热力图、线图、用于关系数据可视化的关系图、Treemap、多维数据可视化的平行坐标，以及用于 BI 的漏斗图、仪表盘，并且支持图与图之间的混搭。

读者可以从 ECharts 官网下载最新的源码包，ECharts 官网如图 9-7 所示。

图 9-7 ECharts 官网

ECharts 的文档齐全，在文档中对各种图形的实例描述很详细，从官方文档入手是再好不过了，读者可以参考它的官网实例和 API 文档。

ECharts 官网实例，请参考图 9-8 所示页面。

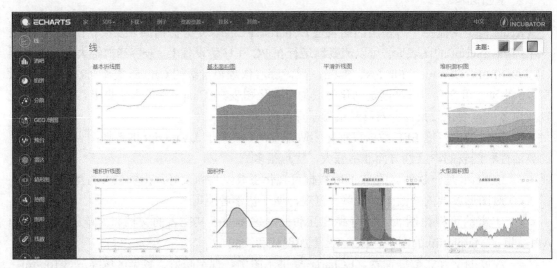

图 9-8　ECharts 官网实例

如果需要定制复杂的图形，可以参考 ECharts API 文档。

首先从 ECharts 官网下载需要的版本。根据用户实现的功能和体积上的需求，ECharts 官网提供了不同打包的下载，如果体积上没有要求，可以直接下载完整版本。本书下载的是 ECharts 完整版本，版本是 4.1.0，文件名是 echarts.min.js，如图 9-9 所示。

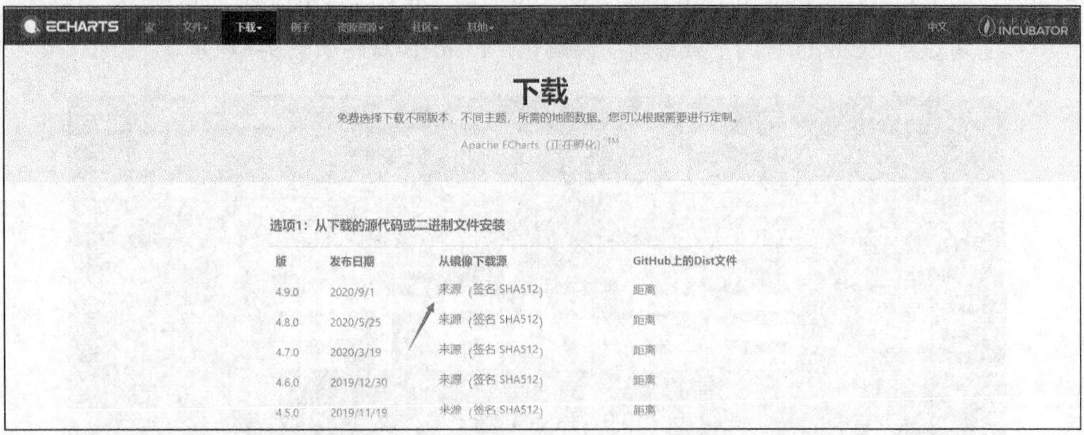

图 9-9　下载 ECharts 完整版本

本实例使用 ECharts 绘制简单的图表。新建 echarts_demo01 目录，在目录下新建文件夹 jslib，把下载的 echarts.min.js 放入此文件夹中。本实例的目录结构如下所示。

```
echarts_demo01
    └── jslib
         └── echarts.min.js
    ├── echarts_demo.html
```

然后在 echarts_demo01 目录下新建 echarts_demo.html 页面。

```
<!DOCTYPE html>
<html>
```

```html
<head>
    <meta charset="utf-8" />
    <title></title>
<!-- 引入 echarts.js -->
<script src="jslib/echarts.min.js"></script>
</head>
<body>
<!-- 为 ECharts 准备一个具备高、宽的 DOM 容器 -->
<div id="main" style="width: 600px;height:400px;"></div>
<script type="text/javascript">
        // 基于准备好的 DOM 容器，初始化 ECharts 实例
        var myChart = echarts.init(document.getElementById('main'));

        // 指定图表的配置项和数据
        option = {
          xAxis: {
            type: 'category',
            data: ['Mon', 'Tue', 'Wed', 'Thu', 'Fri', 'Sat', 'Sun']
            },
           yAxis: {
            type: 'value'
            },
           series: [{
            data: [820, 932, 901, 934, 1290, 1330, 1320],
            type: 'line'
           }]
        };
        // 使用刚指定的配置项和数据显示图表
        myChart.setOption(option);
</script>
</body>
</html>
```

使用浏览器打开 echarts_demo.html 页面，显示效果如图 9-10 所示。

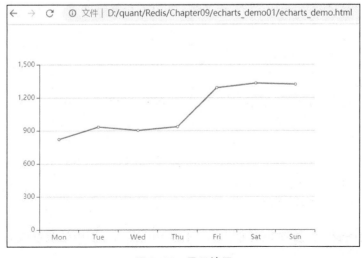

图 9-10　显示效果

本实例中，我们使用 ECharts 绘制简单的曲线图，有以下知识点需要注意。

1. 引入 ECharts

ECharts 是一个纯 JavaScript 的图表库，需要像普通的 JavaScript 库一样用<script>标签引入。

```html
<!DOCTYPE html>
<html>
<head>
<meta charset="utf-8">
<!-- 引入 ECharts 文件 -->
<script src="jslib/echarts.min.js"></script>
</head>
</html>
```

2. 绘制简单图表

在绘制简单图表前需要为 ECharts 准备一个具备高、宽的 DOM 容器。

```html
<body>
<!-- 为 ECharts 准备一个具备高、宽的 DOM 容器 -->
<div id="main" style="width: 600px;height:400px;"></div>
</body>
```

然后就可以通过 echarts.init()方法初始化一个 ECharts 实例并通过 setOption()方法生成一个简单的曲线图。

```html
<script type="text/javascript">
        // 基于准备好的 DOM 容器，初始化 ECharts 实例
        var myChart = echarts.init(document.getElementById('main'));

        // 指定图表的配置项和数据
        option = {
          xAxis: {
            type: 'category',
            data: ['Mon', 'Tue', 'Wed', 'Thu', 'Fri', 'Sat', 'Sun']
          },
          yAxis: {
            type: 'value'
          },
          series: [{
            data: [820, 932, 901, 934, 1290, 1330, 1320],
            type: 'line'
          }]
        };

        // 使用刚指定的配置项和数据显示图表
        myChart.setOption(option);
</script>
```

本实例中的图表的配置项参数可以参考 ECharts 官网文档中的配置项手册，其他图形的配置项也是如此。

本实例使用 ECharts 绘制了一个简单的曲线图,没有使用任何图片,显示的曲线图是通过 ECharts 的 JavaScript 函数绘制的,曲线图中的数据和互动也都是通过 JavaScript 函数实现的,这些 JavaScript 函数实现细节都封装在 ECharts 图表库里。

本实例中绘制的曲线图虽然在显示上是静止的,但在浏览器中是使用 ECharts 的 JavaScript 函数绘制的,支持各种动态互动功能,读者可以好好体会一下。

9.2.2 WebSocket 与消息推送

Redis 自定义监控平台后端获得的 Redis 监控信息是通过 WebSocket 技术推送到前台页面的。WebSocket 是 HTML5 提供的一种浏览器与服务器间进行全双工通信的网络技术,依靠这种技术可以实现客户端和服务器的长连接,实现双向实时通信,如图 9-11 所示。

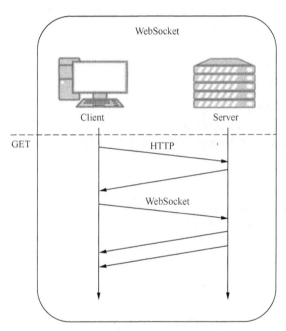

图 9-11　WebSocket 与消息推送

在客户端中,没有必要为 WebSocket 使用 JavaScript 库。实现 WebSocket 的 Web 浏览器将通过 WebSocket 对象公开所有必需的客户端功能(主要指支持 HTML5 的浏览器)。

以下 API 用于创建 WebSocket 对象。

var Socket = new WebSocket(url, [protocol]);

以上代码中的第一个参数 url,指定连接的 URL;第二个参数 protocol 是可选的,指定可接受的子协议。

当创建了 WebSocket 对象后,可以使用 WebSocket 对象的相关事件,如表 9-2 所示。

表 9-2　　　　　　　　　　WebSocket 对象的相关事件

事件	事件处理程序	描述
open	Socket.onopen	连接建立时触发

续表

事件	事件处理程序	描述
message	Socket.onmessage	客户端接收服务器数据时触发
error	Socket.onerror	通信发生错误时触发
close	Socket.onclose	连接关闭时触发

监控页面 index.html 的 WebSocket 核心代码如下。

```
<script type="text/javascript">
    var websocket = null;
    //判断当前浏览器是否支持 WebSocket
    if ('WebSocket' in window) {
        websocket = new WebSocket("ws://localhost:8080/monitor/websocket");
        //console.log('websocket=' + websocket);
    } else {
        alert('当前浏览器 Not support WebSocket')
    }

    //连接发生错误的回调方法
    websocket.onerror = function() {
        setMessageInnerHTML("WebSocket 连接发生错误");
    };

    //连接成功建立的回调方法
    websocket.onopen = function() {
        setMessageInnerHTML("WebSocket 连接成功");
        send();
    }

    //接收到消息的回调方法
    websocket.onmessage = function(event) {
        jsonObj = JSON.parse(event.data);
        //console.log(jsonObj);
        renderBoard(jsonObj);
        renderChat(jsonObj);
    }

    function trim(x) {
        return x.replace(/^\s+|\s+$/gm, '');
    }

    //连接关闭的回调方法
    websocket.onclose = function() {
        setMessageInnerHTML("WebSocket 连接关闭");
    }
```

//监听窗口关闭事件。当窗口关闭时,主动去关闭 WebSocket 连接,防止连接还没断开就关闭窗口,服务器会抛出异常

```javascript
window.onbeforeunload = function() {
    closeWebSocket();
}

//将消息显示在网页上
function setMessageInnerHTML(innerHTML) {
    document.getElementById('message').innerHTML += innerHTML + '<br/>';
}

//关闭 WebSocket 连接
function closeWebSocket() {
    websocket.close();
}

//发送消息
function send() {
    var message = Math.floor(100000 * Math.random());
    websocket.send(message);
}
</script>
```

WebSocket 与 Java 后台交互将返回以下 JSON 对象格式的数据，jsonObj 对象如下。

```
{
    "Stats_total_commands_processed": "416",
    "CPU_used_cpu_sys": "0.30",
    "CPU_used_cpu_user": "0.27",
    "Clients_connected_clients": "2",
    "Memory_used_memory": "0.68",
    "Memory_used_memory_rss": "0.64",
    "time": "2019-10-07 17:10:27",
    "db0_keys": "2"
}
```

返回的数据指标及其含义如表 9-3 所示。

表 9-3 数据指标及其含义

数据指标	含义
Stats_total_commands_processed	运行以来执行过的命令的总数量
CPU_used_cpu_sys	内核空间占用 CPU 的百分比
CPU_used_cpu_user	用户空间占用 CPU 的百分比
Clients_connected_clients	连接客户端的数量
Memory_used_memory	使用总内存
Memory_used_memory_rss	使用物理内存
time	当前的系统时间戳
db0_keys	Redis 数据库 db0 的 key 数量

使用 renderBoard(jsonObj) 函数渲染页面上显示的网格，如图 9-12 所示。

图 9-12　网格

renderBoard(jsonObj)函数的核心代码如下。

```html
<script type="text/javascript">
 function renderBoard(jsonObj){
     //console.log(jsonObj);
     Memory_used_memory = jsonObj.Memory_used_memory;
     Clients_connected_clients = jsonObj.Clients_connected_clients ;
     Memory_used_memory_rss = jsonObj.Memory_used_memory_rss;
     Stats_total_commands_processed = jsonObj.Stats_total_commands_processed;
     CPU_used_cpu_sys = jsonObj.CPU_used_cpu_sys;
     CPU_used_cpu_user = jsonObj.CPU_used_cpu_user;

     document.getElementById('Memory_used_memory').innerHTML = Memory_used_memory + " M";
     document.getElementById('Clients_connected_clients').innerHTML = Clients_connected_clients ;
     document.getElementById('Memory_used_memory_rss').innerHTML = Memory_used_memory_rss + " M" ;
     document.getElementById('Stats_total_commands_processed').innerHTML = Stats_total_commands_processed ;
     document.getElementById('CPU_used_cpu_sys').innerHTML = CPU_used_cpu_sys ;
     document.getElementById('CPU_used_cpu_user').innerHTML = CPU_used_cpu_user ;
 }
</script>
```

使用 renderChat(jsonObj)函数渲染页面上显示的动态曲线图形，如图 9-13 所示。

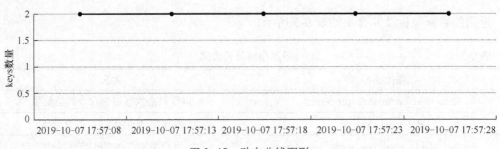

图 9-13　动态曲线图形

使用 ECharts 绘制动态曲线图形的核心代码如下。

首先，在绘图前需要为 ECharts 准备一个有高和宽的 DOM 容器。

```html
<script src="https://cdn.bootcss.com/echarts/3.7.2/echarts.min.js"></script>

<div id="main" style="width: 100%; height: 400px;"></div>
```

然后，通过 echarts.init()方法初始化一个 ECharts 实例并通过 setOption()方法生成一个动态的曲线图。为了实现动态曲线图的效果，曲线图最多只显示 5 个元素，交互数据时删除数组中第一个元素。

```html
<script type="text/javascript">
```
```javascript
    var time_arr = [];
    var data_arr =[];

    function initChat(){
        // 基于准备好的DOM容器,初始化ECharts实例
        var myChart = echarts.init(document.getElementById('main'));

        // 指定图表的配置项和数据
        option = {
                title : {
                    text : 'keys 数量'
                },
                tooltip : {},
                xAxis: {
                   type: 'category',
                   data: time_arr
                },
                yAxis: {
                   type: 'value'
                },
                series: [{
                   data: data_arr,
                   type: 'line',
                   smooth: true
                }]
            };


        // 使用刚指定的配置项和数据显示图表
        myChart.setOption(option);

    }
    function renderChat(jsonObj){
        keys = jsonObj.db0_keys;
        time = jsonObj.time;
        //console.log(keys, time);

        // 使曲线图中最多显示5个元素
        if (time_arr.length < 5) {
           time_arr.push(time);
           data_arr.push(keys);
        }else{
           // 删除数组中的第一个元素
           time_arr.splice(0, 1);
           data_arr.splice(0, 1);

           time_arr[time_arr.length] = time;
           data_arr[data_arr.length] = keys;
```

```
        }

        //console.log(time_arr.length);
        initChat();
    }

    initChat();
</script>
```

9.2.3 创建控制器类

创建控制器类 WebSocketController,调用监控类 MonitorService 获得 Redis 的性能指标。本实例使用"WebSocketController.java",内容如下。

```
package com.xinping.RedisMonitor.server;

import com.xinping.RedisMonitor.service.RedisServiceHandler;
import org.slf4j.Logger;
import org.slf4j.LoggerFactory;
import org.springframework.stereotype.Component;
import javax.websocket.*;
import javax.websocket.server.ServerEndpoint;
import java.io.IOException;
import java.util.concurrent.*;
import java.util.concurrent.atomic.AtomicInteger;

@ServerEndpoint("/websocket")
@Component
public class WebSocketServer {

    private static final Logger log = LoggerFactory.getLogger(WebSocketServer.class);

    /**
     * 当前在线连接数
     */
    private static AtomicInteger onlineCount = new AtomicInteger(0);

    /**
     * 使用CopyOnWriteArraySet对象,concurrent包的线程安全Set,来存放每个客户端对应的
WebSocketServer对象。若要实现服务器与单一客户端通信的话,可以使用Map来存放,其中Key可以为用户标识
     */
    private static CopyOnWriteArraySet<WebSocketServer> webSocketSet = new
CopyOnWriteArraySet<WebSocketServer>();

    /**
     * 与某个客户端的连接会话,需要通过它来给客户端发送数据
     */
    private Session session;

    private Thread queryThread = null;

    /**
```

```
 * 连接建立成功调用的方法
 */
@OnOpen
public void onOpen(Session session) {
    this.session = session;

    webSocketSet.add(this); // 加入Set中
    addOnlineCount(); // 在线连接数加1
    log.info("有新连接加入！当前在线连接数为" + getOnlineCount());
}

/**
 * 连接关闭调用的方法
 */
@OnClose
public void onClose() {
    webSocketSet.remove(this); // 从Set中删除
    subOnlineCount(); // 在线连接数减1
    log.info("有连接关闭！当前在线连接数为" + getOnlineCount());
}

public void closeQueryThread() {
    if (null != queryThread) {
        queryThread.interrupt();
    }
}

/**
 * 收到客户端消息后调用的方法
 *
 * @param message 客户端发送过来的消息
 */
@OnMessage
public void onMessage(String message, Session session) {
    log.info("来自客户端的消息:" + message);

    RedisServiceHandler handler = new RedisServiceHandler(this);
    queryThread = new Thread(handler);
    queryThread.start();
}

/**
 * 发生错误时调用的方法
 *
 * @param session
 * @param error
 */
@OnError
public void onError(Session session, Throwable error) {
    log.error("发生错误原因:" + error.getMessage());
    error.printStackTrace();
```

}

/**
 * 实现服务器主动推送
 */
public void sendMessage(String message) throws IOException {
   if (this.session.isOpen())
       this.session.getBasicRemote().sendText(message);
}

public static synchronized AtomicInteger getOnlineCount() {
   return onlineCount;
}

/**
 * 在线连接数加 1
 */
public static synchronized void addOnlineCount() {
   WebSocketServer.onlineCount.getAndIncrement();
}

/**
 * 在线连接数减 1
 */
public static synchronized void subOnlineCount() {
   WebSocketServer.onlineCount.getAndDecrement();
}

}
```

@ServerEndpoint("/websocket")注解是一个类层次的注解，它的功能主要是将目前的类定义成一个 WebSocket 服务器，注解的值将被用于监听用户连接的终端访问的 URL 地址，客户端可以通过这个 URL 地址来连接 WebSocket 服务器。本例创建的项目名称是 monitor，部署在端口为 8080 的 Tomcat 服务器上，所以本例中 WebSocket 对象的访问地址是 ws://localhost:8080/monitor/websocket，对应的 HTML5 的 WebSocket 核心代码如下。

```
var websocket = null;
//判断当前浏览器是否支持 WebSocket
if ('WebSocket' in window) {
   websocket = new WebSocket("ws://localhost:8080/monitor/websocket");
   //console.log('websocket=' + websocket);
} else {
   alert('当前浏览器 Not support WebSocket')
}
```

9.2.4　业务逻辑

创建监控类 MonitorService，用来监控 Redis 状态，调用 INFO 命令，解析结果，并对结果进行转换处理。本实例使用了 "MonitorService.java"，内容如下。

```
package com.dxtd.monitor.service;
```
```

```java
import java.util.HashMap;
import java.util.Map;
import javax.tools.Tool;
import com.dxtd.monitor.util.RedisUtil;
import com.dxtd.monitor.util.Tools;
import redis.clients.jedis.Jedis;

public class MonitorService {
    //调用INFO命令
    public String getInfo() {
        String infoContent = null;
        Jedis jedis = null;
        try {
            jedis = RedisUtil.getJedis();
            infoContent = jedis.info();
        } catch (Exception e) {
            e.printStackTrace();
        } finally {
            if (jedis != null) {
                jedis.close();
            }
        }
        return infoContent;
    }

    public Double getIntValue(Map<String, String> infoMap, String key) {
        if (null == infoMap)
            return 0.0;

        Double value = Double.valueOf(infoMap.get(key));
        return value;
    }

    public String getStringValue(Map<String, String> infoMap, String key) {
        if (null == infoMap)
            return "";

        return infoMap.get(key).trim();
    }

    public Integer getKeys(String info, String key) {
        Integer keysNumValue = null;
        String[] strs = info.split("\n");
        if (strs != null && strs.length > 0) {
            for (int i = 0; i < strs.length; i++) {
                String s = strs[i].trim();

                if (s.indexOf(key) > -1) {
                    String[] str = s.split(",");
                    if (null != str) {
                        String[] dbs = str[0].split(":");
```

```java
                            String[] dbKeys = dbs[1].split("=");
                            String keyStr = dbKeys[0];
                            keysNumValue = Integer.valueOf(dbKeys[1]);
                            break;
                        }
                    }else {
                        keysNumValue = 0;
                    }
                }
            }
            return keysNumValue;
        }

        //解析结果
        public Map parseInfo(String content) {
            String[] lines = content.split("\n");
            Map infoMap = new HashMap();
            String part = null;
            for (String line : lines) {
                if (line.isEmpty()) {
                    continue;
                }

                if (line.startsWith("#")) {
                    part = line.replace("#", "").trim();
                    continue;
                }

                int index = line.indexOf(':');
                if (index >= 0) {
                    infoMap.put(part + "_" + line.substring(0, index), line.substring(index + 1));
                }
            }

            return infoMap;
        }

        //数据转换处理
        public void transData(String infoContent) {
            Map<String, String> infoMap = parseInfo(infoContent);
            // 内核空间占用 CPU 的百分比
            String ucs = getStringValue(infoMap, "CPU_used_cpu_sys");
            // 用户空间占用 CPU 的百分比
            String ucu = getStringValue(infoMap, "CPU_used_cpu_user");
            // 阻塞客户端的数量
            String cbc = getStringValue(infoMap, "Clients_blocked_clients");
            // 连接客户端的数量
            String ccc = getStringValue(infoMap, "Clients_connected_clients");
            // 使用总内存
            String mum = getStringValue(infoMap, "Memory_used_memory");
```

```java
        // 使用物理内存
        String mur = getStringValue(infoMap, "Memory_used_memory_rss");
        // 运行以来执行过的命令的总数量
        String cmd = getStringValue(infoMap, "Stats_total_commands_processed");
        // 每秒过期 key 数量
        String exp = getStringValue(infoMap, "Stats_expired_keys");
        // 每秒淘汰 key 数量
        String evt = getStringValue(infoMap, "Stats_evicted_keys");
        // 每秒命中数量
        String hit = getStringValue(infoMap, "Stats_keyspace_hits");
        // 每秒丢失数量
        String mis = getStringValue(infoMap, "Stats_keyspace_misses");

        Integer db0keysNum = getKeys(infoContent, "db0_keys");

        long thisTs = System.currentTimeMillis();

        System.out.println("ucs=" + ucs + ",ucu=" + ucu + ",cbc=" + cbc);
        System.out.println("ccc=" + ccc + ",mum=" + mum + ",mur=" + mur);
        System.out.println("cmd=" + cmd + ",exp=" + exp);
        System.out.println("evt=" + evt + ",hit=" + hit + ",mis=" + mis);
        System.out.println("db0keysNum=" + db0keysNum);
    }

    public Map<String, Object> getRedisInfo(String infoContent) {
        Map<String, Object> map = new HashMap<String, Object>();
        Map<String, String> infoMap = parseInfo(infoContent);
        // 使用总内存
        String mum = getStringValue(infoMap, "Memory_used_memory");
        // 连接客户端的数量
        String ccc = getStringValue(infoMap, "Clients_connected_clients");
        // 使用物理内存
        String mur = getStringValue(infoMap, "Memory_used_memory_rss");
        // 运行以来执行过的命令的总数量
        String cmd = getStringValue(infoMap, "Stats_total_commands_processed");
        // 内核空间占用 CPU 的百分比
        String ucs = getStringValue(infoMap, "CPU_used_cpu_sys");
        // 用户空间占用 CPU 的百分比
        String ucu = getStringValue(infoMap, "CPU_used_cpu_user");
        Integer db0keysNum = getKeys(infoContent, "db0:keys");

        map.put("Memory_used_memory", Tools.transByteToMBSize(Integer.valueOf(mum )) );
        map.put("Clients_connected_clients", ccc);
        map.put("Memory_used_memory_rss", Tools.transByteToMBSize(Integer.valueOf(mur )) );
        map.put("Stats_total_commands_processed", cmd);
        map.put("CPU_used_cpu_sys", ucs);
        map.put("CPU_used_cpu_user", ucu);
        map.put("db0_keys", db0keysNum+"");
        map.put("time", Tools.getCurrntTime());
        return map;
```

        }

    }

创建 RedisServiceHandler 类实现一个线程,在这个线程的 run()方法里,生成 MonitorService 对象来监控 Redis 的状态,并对 INFO 命令的返回值进行处理,然后把返回结果返还给 WebSocket 的前端页面。本实例使用"RedisServiceHandler.java",内容如下。

```
package com.dxtd.monitor.service;

import java.io.IOException;
import java.util.Map;

import com.alibaba.fastjson.JSON;
import com.dxtd.monitor.action.WebSocketController;

public class RedisServiceHandler implements Runnable {

 private WebSocketController webSocket;

 public RedisServiceHandler(WebSocketController webSocket){
     this.webSocket = webSocket;
 }

 ublic void run() {
     while(true){
         try {
             MonitorService monitor = new MonitorService();
             String info = monitor.getInfo();
             Map<String,Object> redisInfo = monitor.getRedisInfo(info);
             //群发消息
             webSocket.sendMessage(JSON.toJSONString(redisInfo) );

             //每 5s 发送一次消息,以便页面更新数据
             Thread.sleep(1000 * 5 );
         } catch (IOException e) {
             e.printStackTrace();
         } catch (InterruptedException e) {
             e.printStackTrace();
         }
     }
 }

}
```

9.2.5 常用工具类

RedisUtil 类是连接 Redis 的工具类,封装了 Jedis 连接池的业务逻辑。本实例使用 "RedisUtil.java",内容如下。

```
package com.dxtd.monitor.util;

import redis.clients.jedis.Client;
```

```java
import redis.clients.jedis.Jedis;
import redis.clients.jedis.JedisPool;
import redis.clients.jedis.JedisPoolConfig;

public class RedisUtil {
    private static JedisPool jedisPool;

    /**
     * 建立 Redis 的连接池，如果需要优化可以把连接池的配置参数单独抽取出来
     *
     */
    private static void createJedisPool() {

        JedisPoolConfig jedisPoolConfig = new JedisPoolConfig();
            //指定连接池中最大空闲连接数
            jedisPoolConfig.setMaxIdle(10);
            //连接池中创建的最大连接数
            jedisPoolConfig.setMaxTotal(100);
            //设置创建连接的超时时间
            jedisPoolConfig.setMaxWaitMillis(2000);
            //表示连接池在创建连接的时候会先测试一下连接是否可用，这样可以保证连接池中的连接都可用
            jedisPoolConfig.setTestOnBorrow(true);

        // 创建连接池
        jedisPool = new JedisPool(jedisPoolConfig, "127.0.0.1", 6379);
    }

    /**
     * 在多线程环境同步初始化
     */
    private static synchronized void poolInit() {
        if (jedisPool == null)
            createJedisPool();
    }

    /**
     * 获取一个 Jedis 对象
     *
     * @return
     */
    static {
        if (jedisPool == null)
            poolInit();
    }

    /**
     * 归还一个连接
     *
     * @param jedis
     */
    public static void returnRes(Jedis jedis) {
```

```java
        jedis.close();
    }

    public static Jedis getJedis() {
        if (jedisPool == null)
            poolInit();

        Jedis jedis = jedisPool.getResource();
        return jedis;
    }

    /**
     * 获取 Redis 服务器信息
     *
     * @param String
     */
    public String getRedisInfo() {
        if (jedisPool == null)
            poolInit();

        Jedis jedis = null;
        try {
            jedis = jedisPool.getResource();
            Client client = jedis.getClient();
            client.info();
            String info = client.getBulkReply();
            return info;
        } finally {
            // 返还到连接池
            if(null != jedis)
                jedis.close();
        }
    }

}
```

常用工具类 Tools，用于获得格式化后的时间和转化单位。本实例使用"Tools.java"，内容如下。

```java
package com.dxtd.monitor.util;

import java.text.DecimalFormat;
import java.text.SimpleDateFormat;
import java.util.Date;

public class Tools {

    //获得格式化后的时间
    public static String getCurrntTime() {
        SimpleDateFormat format = new SimpleDateFormat("yyyy-MM-dd HH:mm:ss");
        Date now = new Date();
        return format.format(now);
```

```java
    }

// 转化单位，从 Byte 到 MB
 public static String transByteToMBSize(int size) {
     int MB = 1024 * 1024;// 定义 1024MB 的计算存储单位
     DecimalFormat df = new DecimalFormat("0.00");// 格式化小数
     String resultSize = "";
     resultSize = df.format(size / (float) MB);
     return resultSize;
 }

}
```

第 10 章 Redis 的缓存设计与优化

本章主要讲解 Redis 的缓存设计与优化，以及在生产环境中遇到的 Redis 常见问题，例如缓存雪崩和缓存穿透，还讲解了相关问题的解决方案。

10.1 Redis 缓存的优点和缺点

使用 Redis 作为缓存有以下优点。

1. 高速读写

Redis 可以帮助解决由于数据库压力造成的延迟现象，针对很少改变的数据并且经常使用的数据，我们可以把这些数据放入内存中。这样一方面可以减小数据库压力，另一方面可以提高读写效率。

2. 降低后端负载

后端服务器通过缓存降低负载，业务端使用 Redis 可以降低后端数据库 MySQL 的负载等。

使用 Redis 作为缓存带来的代价有以下几点。

1. 数据不一致

程序的缓存层和数据层有时会不一致，这和更新数据策略有关。

2. 代码维护成本

原本只需要读写 MySQL 就能实现功能，但加入了 Redis 缓存之后就要去维护缓存中的数据，增加了代码复杂度。

3. 堆内缓存可能带来内存溢出的风险，从而影响用户进程

在 Java 虚拟机的 EhCache、LoadingCache、Java 虚拟机栈、方法区、本地方法栈、程序计数器中，堆内缓存可能会带来内存溢出的风险，从而影响用户进程。

10.2 缓存雪崩

1. 什么是缓存雪崩

缓存雪崩是指数据未加载到缓存中,或者缓存在同一时间大面积失效,导致所有请求都查询数据库,从而导致数据库 CPU 和内存负载过高,甚至数据库宕机。

2. 有什么解决方案来防止缓存雪崩

(1)使用互斥锁(mutex)。使用互斥锁来防止缓存雪崩,使用 Redis 的 SETNX 命令去设置一个 mutex key,当操作返回成功时,再执行查询数据库操作并回设 Redis 缓存。否则,就重试执行缓存的 GET 方法。

(2)缓存预热。缓存预热就是应用上线后,将相关的缓存数据直接加载到缓存系统中。这样用户就可以直接查询事先被预热的缓存数据。

(3)双层缓存策略。Cache 1 为原始缓存,Cache 2 为复制缓存。Cache 1 失效时,可以访问 Cache 2。Cache 1 缓存失效时间设置为短期,Cache 2 缓存失效时间设置为长期。

(4)定时更新缓存策略。对失效性要求不高的缓存,在容器启动初始化加载时采用定时任务更新或移除缓存。

(5)设置不同的过期时间,让缓存失效的时间点尽量均匀。

10.3 缓存穿透

1. 什么是缓存穿透

缓存就是数据交换的缓冲区。缓存的主要作用是提高查询效率。在企业开发的软件系统中常常使用 Redis 作为缓存中间件,当请求到达服务器端时,优先查询缓存中的数据,当缓存中不存在时,再查询数据库,如果在数据库中查询到数据会将数据写回缓存,使得下一次同样的请求能够在缓存中直接查询到数据。一些攻击性请求会特意查询缓存中不存在的数据,产生缓存穿透。

缓存穿透是指查询一个不存在的数据。例如,Redis 在缓存中没有查询到要查询的数据,需要去数据库查询,如果查询不到数据则不写入缓存,这将导致这个不存在的数据在每次请求时都到数据库查询,进而对数据库产生流量冲击造成缓存穿透。

2. 有什么解决方案来防止缓存穿透

(1)采用布隆过滤器。有关布隆过滤器的知识在下节详细介绍。

(2)缓存空值。如果一个查询返回的数据为空值,那么不管是数据不存在,还是系统故障,程序仍然会把这个空值进行缓存处理,但它的过期时间会很短,可能不超过 5min。通过设置的默认值将该数据直接存放到缓存中,这样第二次在缓存中就可以查询到值了,而不会继续访问数据库。

10.4 布隆过滤器

10.4.1 布隆过滤器简介

布隆过滤器（Bloom Filter）是 1970 年由布隆提出的。它实际上是一个很长的二进制向量和一系列随机映射函数。布隆过滤器可以用于检索一个元素是否在一个集合中。它的优点是空间效率高和查询时间短，缺点是有一定的误识别率和元素删除困难。

布隆过滤器是一种空间效率很高的随机数据结构，它利用位数组很简洁地表示一个集合，并能判断一个元素是否属于这个集合。布隆过滤器由一个很长的位数组和一系列散列函数组成，数组的每个元素都只占 1 bit 空间，并且每个元素只能为 0 或 1。

布隆过滤器还拥有 k 个散列函数，当一个元素加入布隆过滤器时，会使用 k 个散列函数对其进行 k 次计算，得到 k 个散列值，并且根据得到的 k 个散列值，在位数组中把对应位置的值置为 1。判断某个元素是否在布隆过滤器中，就对该元素进行 k 次散列计算，判断得到的值在位数组中对应位置的值是否都为 1，如果每个元素都为 1，就说明这个元素在布隆过滤器中。

将数据库中需要查询的数据放入系统缓存中的布隆过滤器中，当请求向后台系统查询数据时，先去系统缓存中的布隆过滤器中进行查找，如果查询的数据在布隆过滤器中不存在，就不用查询数据库了，直接给请求返回一个未查询到数据的结果，从而避免了对数据库的频繁查询。

布隆过滤器是一个判断元素是否属于集合的快速的概率算法。布隆过滤器有可能会出现错误判断，但不会漏掉判断。也就是说，如果布隆过滤器判断元素不在集合中，那么肯定不在；如果判断元素在集合中，那么会有一定的概率判断错误。因此，布隆过滤器不适合那些零错误的应用场景。而在能容忍低错误率的应用场景中，布隆过滤器比其他常见的算法（如散列函数、折半查找）极大节省了空间，如图 10-1 所示。

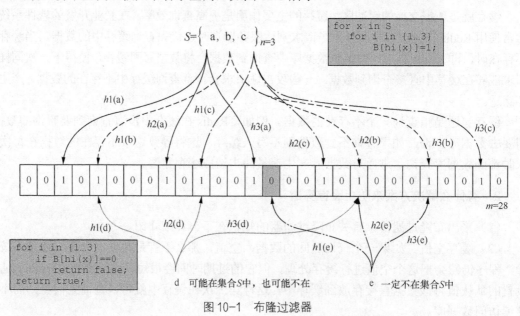

图 10-1　布隆过滤器

布隆过滤器很常用的一个功能是去重，比如在爬虫中有一个常见的需求：目标网站的 URL 可以有成千上万个，怎么判断某个 URL 是否被爬虫爬取过呢？一个简单的方法是，可以把爬虫爬取过的每个 URL 存入数据库中，每次一个新的 URL 过来就到数据库查询是否爬取过。例如，SELECT * FROM spider WHERE url = 'http://www.163.com'。

但是随着爬虫爬取过的 URL 越来越多，每次请求查询时都要访问数据库一次，判断某个 URL 是否访问过使用 SQL 查询效率并不高。除了数据库之外，还可以使用 Redis 的 Set 数据类型满足这个需求，并且其性能优于数据库。但是 Redis 也存在一个问题，它会耗费过多的内存，这时候就可以使用布隆过滤器来解决去重问题。相比于数据库和 Redis，使用布隆过滤器可以很好地避免性能和内存占用的问题。

我们通常使用 Redis 作为数据缓存，当收到请求时先通过 key 去 Redis 缓存中查询，如果查询的数据在 Redis 缓存中不存在，就会去查询数据库中的数据。如果这种请求量很大，会给数据库造成很大的查询压力，从而影响系统的性能，这时就需要用到布隆过滤器来解决缓存穿透问题了。

解决缓存穿透的方法。

方法一：当数据库和 Redis 中都不存在 key，查询数据库会返回 null。需要在 Redis 中使用 SETEX key null expireTime 设置一个过期时间 expireTime，这样当再次请求 key 时 Redis 将直接返回 null，而不用再次查询数据库。

方法二：使用 Redis 提供的布隆过滤器模块 RedisBloom，同样是将存在的 key 放入布隆过滤器中。当收到请求时先在布隆过滤器中查询 key 是否存在，如果 key 不存在直接返回 null，不必再次查询数据库。

布隆过滤器的用途是判断过滤器中是否存在该数据，从而减少没有必要的数据库请求。

10.4.2　Redis 加载布隆过滤器模块

Redis 官方提供的布隆过滤器在 Redis 4.0 发布以后才正式推出。布隆过滤器可作为一个插件加载到 Redis 服务器中，给 Redis 提供强大的布隆去重功能。在本小节中，我们将学习如何在 Redis 服务器上加载布隆过滤器模块。

在 GitHub 搜索 RedisBloom 下载最新发布的源代码，单击页面的"Clone or download"按钮后选择"Download ZIP"，下载 RedisBloom-master.zip 到本地硬盘，如图 10-2 所示。

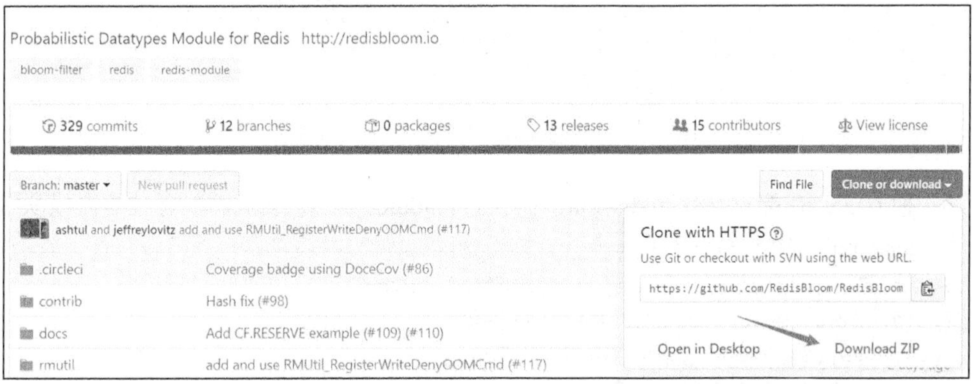

图 10-2　下载最新发布的源代码

上传 RedisBloom-master.zip 到 Linux 服务器，在 Linux 服务器上进行解压缩和编译。

```
$ unzip RedisBloom-master.zip
$ cd RedisBloom-master/
$ make
```

编译后得到动态库 redisbloom.so，如图 10-3 所示。

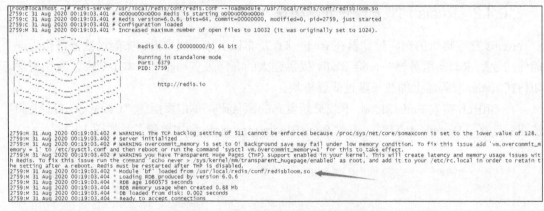

图 10-3　得到动态库 redisbloom.so

复制动态库 redisbloom.so 到 /usr/local/redis/conf 目录下。

```
$ cp redisbloom.so /usr/local/redis/conf
```

启动 Redis 服务器加载布隆过滤器模块主要有以下两种方法。

第一种方法：启动 Redis 服务器时，使用参数 --loadmodule 就可以加载布隆过滤器模块。

```
$ redis-server /usr/local/redis/conf/redis.conf --loadmodule /usr/local/redis/conf/redisbloom.so
```

从图 10-4 可以看出，在 Redis 服务器启动时加载了布隆过滤器模块。

图 10-4　启动 Redis 加载布隆过滤器模块

第二种方法：在配置文件 redis.conf 中通过 loadmodule 命令加载布隆过滤器模块。在配置文件 redis.conf 中加上如下内容。

```
loadmodule /usr/local/redis/conf/redisbloom.so
```

然后使用 redis-server 命令启动 Redis 服务器。

```
$ redis-server /usr/local/redis/conf/redis.conf
```

启动 Redis 服务器后的消息如图 10-5 所示。

```
3081:M 25 Oct 19:48:24.584 # WARNING: The TCP backlog setting of 511 cannot be enforced because /proc/sys/net/core
/somaxconn is set to the lower value of 128.
3081:M 25 Oct 19:48:24.584 # Server initialized
3081:M 25 Oct 19:48:24.584 # WARNING overcommit_memory is set to 0! Background save may fail under low memory condi
tion. To fix this issue add 'vm.overcommit_memory = 1' to /etc/sysctl.conf and then reboot or run the command 'sy
sctl vm.overcommit_memory=1' for this to take effect.
3081:M 25 Oct 19:48:24.585 # WARNING you have Transparent Huge Pages (THP) support enabled in your kernel. This wi
ll create latency and memory usage issues with Redis. To fix this issue run the command 'echo never > /sys/kernel/
mm/transparent_hugepage/enabled' as root, and add it to your /etc/rc.local in order to retain the setting after a
reboot. Redis must be restarted after THP is disabled.
3081:M 25 Oct 19:48:24.585 * Module 'bf' loaded from /usr/local/redis/conf/redisbloom.so
3081:M 25 Oct 19:48:24.585 * Ready to accept connections
```

图 10-5　启动 Redis 服务器后的消息

从图 10-5 可以看出，在 Redis 服务器启动时加载了布隆过滤器模块。

1．布隆过滤器命令

布隆过滤器命令的基本语法如下。

```
BF.RESERVE {key} {error_rate} { initial_size }
```

参数说明如下。

- error_rate 指允许布隆过滤器的错误率（容错率），取值范围为 0～1。数值越小，占用内存越大，操作时占用的 CPU 资源越多。
- initial_size 指布隆过滤器的容量，当实际存储的元素个数超过 initial_size 后，布隆过滤器的准确率就会下降。

默认的容错率 error_rate 是 0.01，容量 initial_size 是 100。如果不通过 BF.RESERVE 命令来新建布隆过滤器，添加元素时就会自动创建布隆过滤器，但会使用默认的容错率与容量。

BF.ADD 用于添加元素到布隆过滤器中。如果布隆过滤器不存在，则会自动创建，使用默认的容错率与容量。

```
BF.ADD {newFilter} {foo}
```

BF.EXISTS 用于判断某个元素是否在布隆过滤器中。检查布隆过滤器中是否存在该元素，不存在则返回 0，存在则返回 1。

```
BF.EXISTS {newFilter} {foo}
```

2．布隆过滤器的实例

使用 BF.RESERVE 命令新建布隆过滤器 urls。

```
127.0.0.1:6379> BF.RESERVE urls 0.01 100
OK
```

使用这个命令要注意一点：使用这个命令之前布隆过滤器的名字应该不存在，如果使用命令之前布隆过滤器的名字已经存在，就会报错"(error) ERR item exists"。

```
# 在布隆过滤器 urls 中加入两条网页数据
127.0.0.1:6379> BF.ADD urls http://www.163.com
(integer) 1
127.0.0.1:6379> BF.ADD urls http://www.cnblogs.com
(integer) 1
# 网页 http://www.cnblogs.com 在布隆过滤器 urls 中存在
127.0.0.1:6379> BF.EXISTS urls http://www.cnblogs.com
(integer) 1
# 网页 http://www.zol.com 在布隆过滤器 urls 中不存在
127.0.0.1:6379> BF.EXISTS urls http://www.zol.com
(integer) 0
```

10.4.3 在项目中使用布隆过滤器

使用 IntelliJ IDEA 新建 Maven 项目,命名为 RedisbloomDemo。本实例目录为 Redis\Chapter10\RedisbloomDemo,内容如下。

在 RedisbloomDemo 项目的 pom.xml 文件中引入以下类库。

```xml
<dependencies>
    <dependency>
        <groupId>com.redislabs</groupId>
        <artifactId>jrebloom</artifactId>
        <version>1.0.1</version>
    </dependency>

    <dependency>
        <groupId>redis.clients</groupId>
        <artifactId>jedis</artifactId>
        <version>3.1.0</version>
    </dependency>
</dependencies>
```

新建测试类 RedisbloomDemo。本实例使用"RedisbloomDemo.java",内容如下。

```java
package com.dxtd.redis;

import io.rebloom.client.Client;

public class RedisbloomDemo {
    public static void main(String[] args) {
        // 创建客户端, Jedis 实例
        Client client = new Client("192.168.11.15", 6379);

        String urlsBloomKey = "urls";

        // 创建一个有初始值和出错率的布隆过滤器
        client.createFilter(urlsBloomKey,1000,0.01);
        // 在布隆过滤器新增一个 key-value 键值对
        boolean url1 = client.add(urlsBloomKey,"http://www.163.com");
        System.out.println("url1 add : " + url1);

        boolean url2 = client.add(urlsBloomKey,"http://www.cnblogs.com");
        System.out.println("url2 add : " + url1);

        // 某个 value 是否在布隆过滤器中存在
        boolean exists = client.exists(urlsBloomKey, "http://www.163.com");
        System.out.println("http://www.163.com 是否存在: " + exists);
    }
}
```

该程序输出如下。

```
url1 add : true
url2 add : true
http://www.163.com 是否存在: true
```

第 11 章 扩展知识

因为本书涉及的技术内容比较多，所以把读者需要掌握的内容单独汇聚成一章，读者可以根据需要阅读自己感兴趣的内容。本章内容是根据笔者的实际开发经验整理的，列出了常用的 CentOS 配置、Maven 配置、IntelliJ IDEA 配置、Chrome 的常用技巧和使用 VMware 虚拟机的常见配置，希望能对读者的工作和学习有所帮助。

11.1 配置 CentOS 7

CentOS 是一个基于 Red Hat Linux 提供的可自由使用源代码的企业级 Linux 发行版。每个版本的 CentOS 都会获得 10 年的技术支持（通过安全更新方式），建立一个安全、低维护、稳定、高预测性和高重复性的 Linux 环境。CentOS 是 Community Enterprise Operating System 的缩写。

本节介绍 CentOS 7 的常用配置方法。

11.1.1 关闭防火墙

使用 FTP 软件从 Windows 向 CentOS 传送文件时，为了传输方便，可以关闭 CentOS 7 上的防火墙。

彻底关闭 CentOS 7 上的防火墙需要经过以下两步。

关闭防火墙步骤 1。

```
$ systemctl stop firewalld.service        # 停止防火墙
$ systemctl disable firewalld.service     # 禁止防火墙开机启动
```

关闭防火墙步骤 2。

使用 vi 命令修改 /etc/sysconfig/selinux 配置文件，把 SELINUX 配置项改为 disabled。

```
$ vi /etc/sysconfig/selinux
```

具体内容如图 11-1 所示。

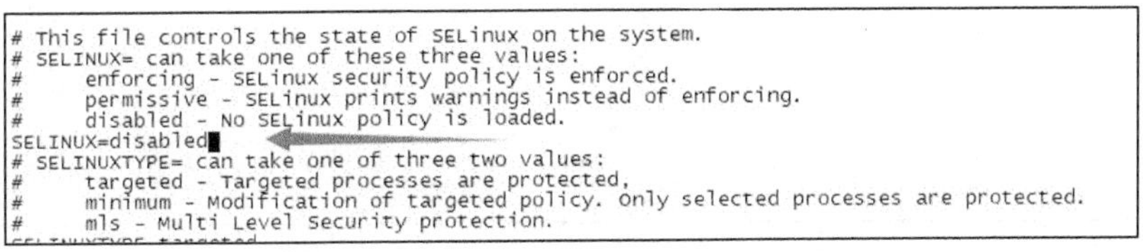

```
# This file controls the state of SELinux on the system.
# SELINUX= can take one of these three values:
#     enforcing - SELinux security policy is enforced.
#     permissive - SELinux prints warnings instead of enforcing.
#     disabled - No SELinux policy is loaded.
SELINUX=disabled
# SELINUXTYPE= can take one of three two values:
#     targeted - Targeted processes are protected,
#     minimum - Modification of targeted policy. Only selected processes are protected.
#     mls - Multi Level Security protection.
```

图 11-1 关闭防火墙

237

输入以下命令彻底关闭 SELinux（Security-Enhanced Linux）。SELinux 是美国国家安全局（NSA）对强制访问控制的实现，是 Linux 历史上杰出的安全子系统。

```
$ setenforce 0
```

修改防火墙配置文件后，最好重启一下 Linux 操作系统。

11.1.2　配置国内 yum 仓库

通常使用 yum install 命令在 Linux 系统中安装的软件，这种方式可以自动处理软件的依赖关系，并且一次安装所有依赖的软件包。一般情况下，从国外的镜像仓库下载软件速度非常慢，很多情况下无法成功下载。因此国内的一些大公司和科研机构做了镜像仓库同步国外的软件，我们可以使用国内的镜像仓库下载 Linux 软件。本书使用清华大学提供的镜像仓库进行 yum 仓库配置。

在网上搜索清华大学镜像仓库，找到镜像列表里的"centos"，如图 11-2 所示。

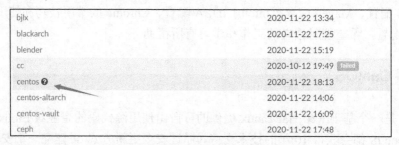

图 11-2　清华大学的镜像仓库列表

单击 centos 后面的"?"，进入帮助说明页面。本书使用的实验环境是在 CentOS 7 下安装和操作 Redis，所以在帮助说明页面中选择 CentOS 版本为"CentOS 的 7"，如图 11-3 所示。

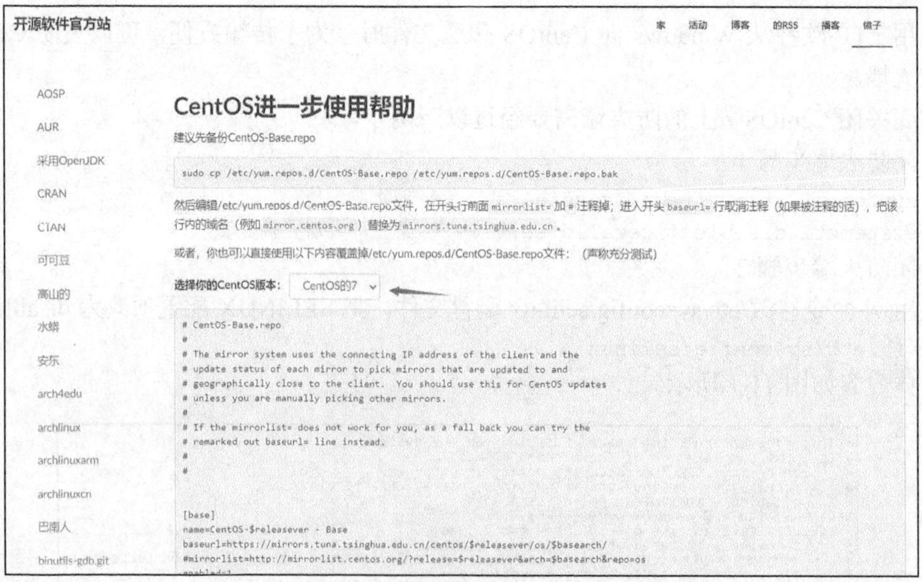

图 11-3　选择 CentOS 版本

1. 首先备份 CentOS-Base.repo 文件

```
$ sudo cp /etc/yum.repos.d/CentOS-Base.repo /etc/yum.repos.d/CentOS-Base.repo.bak
```

2. 将清华大学镜像仓库信息写入 /etc/yum.repos.d/CentOS-Base.repo

使用 vi 命令编辑 CentOS-Base.repo 文件

```
$ vi /etc/yum.repos.d/CentOS-Base.repo
```

将帮助说明页面中"选择你的 CentOS 版本"下面的内容进行复制，粘贴到 CentOS-Base.repo 文件中，如图 11-4 所示。

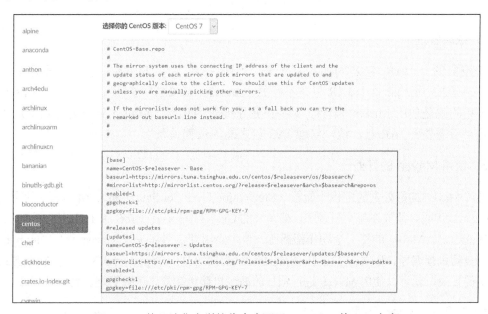

图 11-4　使用清华大学镜像仓库配置 CentOS 7 的 yum 仓库

3. 清除缓存

```
$ yum clean all       # 清除系统所有的 yum 缓存
$ yum makecache       # 生成 yum 缓存
```

4. 更新系统

```
yum -y update
```

11.2　Maven 基础知识

Maven 是通过 pom.xml 描述信息来管理项目的构建、报告和文档的项目管理工具。本节主要介绍 Maven 基础知识。

11.2.1　Maven 的基本概念

Maven（翻译为"专家""内行"）是跨平台的项目管理工具，主要服务基于 Java 的项目构建、依赖管理。

1. 项目构建

项目构建过程包括清理、编译、测试、报告、打包、部署这 6 个步骤，这 6 个步骤就是一个项目的完整构建过程，如图 11-5 所示。

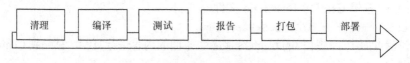

图 11-5　项目的完整构建过程

2. 依赖管理

依赖指的是 JAR 包之间的相互依赖，例如搭建一个 Struts2 的开发框架时，只有 struts2-core-2.3.16.3.jar 这个 JAR 包是不行的，struts2-core-2.3.16.3.jar 还依赖其他 JAR 包，依赖管理指的就是使用 Maven 来管理项目中使用到的 JAR 包，Maven 管理的方式就是"自动下载项目所需要的 JAR 包，统一管理 JAR 包之间的依赖关系"。

3. 使用 Maven 的好处

使用 Maven 的好处是约定优于配置（Convention Over Configuration）。Maven 中使用约定，约定 Java 源代码必须放在哪个目录下，编译好的 Class 代码又必须放在哪个目录下，这些目录都有明确的约定。Maven 的每一个动作都拥有一个生命周期，例如执行 Maven 命令"mvn install"就可以自动执行编译、测试、打包等。构建过程只需要定义一个 pom.xml 文件，然后把源代码放到默认的目录，就可以使用 Maven 进行项目的高度自动化构建、依赖管理、仓库管理。

11.2.2　Maven 下载

在 Apache 官网下载 Maven 的压缩包，如图 11-6 所示。

图 11-6　下载 Maven 的压缩包

下载完成后，得到一个压缩包 apache-maven-3.6.3-bin.zip，解压缩后可以看到 Maven 的目录结构，如图 11-7 所示。

Maven 的目录结构说明如下。
- bin：含有 mvn 运行的脚本。
- boot：含有 plexus-classworlds 类加载器框架。

- conf：含有 settings.xml 配置文件。
- lib：含有 Maven 运行时所需要的 Java 类库。

图 11-7　Maven 的目录结构

11.2.3　Maven 安装

（1）首先要确保计算机上已经安装了 JDK（JDK 1.8 以上的版本），配置好 JDK 的系统变量，使用如下两个命令检查 JDK 安装的情况，如图 11-8 所示。

```
echo %JAVA_HOME%
java -version
```

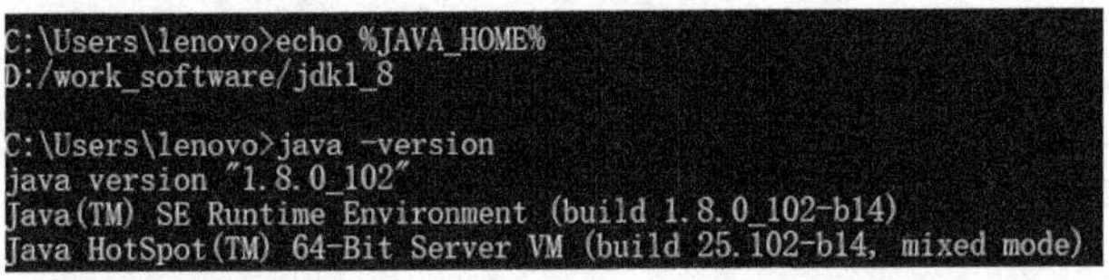

图 11-8　检查 JDK 安装的情况

（2）对 apache-maven-3.6.3-bin.zip 进行解压缩，解压缩的目录不要有中文，例如解压缩到 D:/work_software 目录下。

（3）配置系统变量 M2_HOME，如图 11-9 所示。

图 11-9　配置系统变量 M2_HOME

Maven 配置是以笔者的开发环境为例，读者需要根据实际情况进行变动。

（4）配置系统变量 Path，将%M2_HOME%/bin 加入 Path 中，一定注意要将其用分号与其他值隔开，如图 11-10 所示。

图 11-10　配置系统变量 Path

（5）查看 Maven 的版本。输入 mvn -v 命令查看 Maven 的版本，如图 11-11 所示。

图 11-11　查看 Maven 的版本

11.2.4　修改从 Maven 中心仓库下载到本地的 JAR 包的默认存储位置

从 Maven 中心仓库下载到本地的 JAR 包默认存储在${user.home}/.m2/repository 中，${user.home}表示当前登录系统的用户目录（例如 C:\Users\lenovo），如图 11-12 和图 11-13 所示。

图 11-12　.m2 目录

图 11-13　Maven 的 repository

以上是 JAR 包的默认存储位置，可以设置自定义下载到本地时的 JAR 包的存放目录。例如在 D:\目录下创建一个 maven_jar 文件夹。

找到%/apache-maven/conf 目录下的 settings.xml 配置文件，如图 11-14 所示。

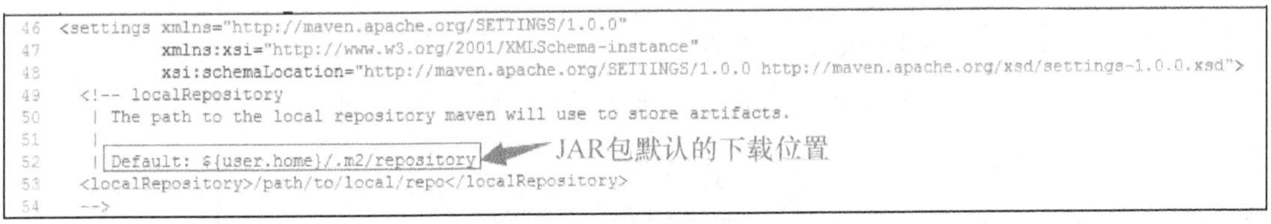

图 11-14　Maven 的 JAR 包默认的下载位置

在 settings.xml 配置文件的<settings></settings>节点里添加以下代码，如图 11-15 所示。

<localRepository>D:/maven_jar</localRepository>

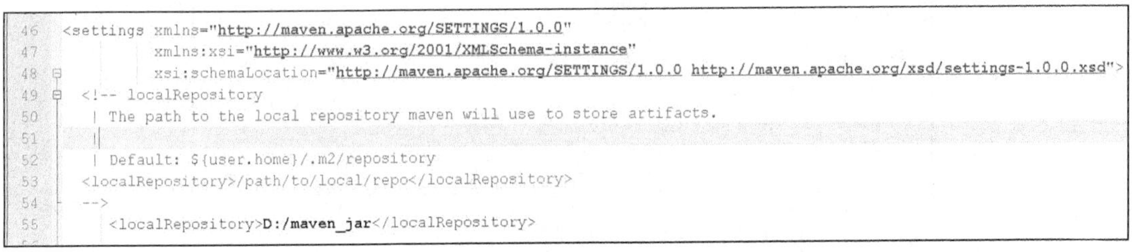

图 11-15　修改 Maven 的 JAR 包下载位置

这样就可以把 JAR 包下载到我们指定的 D:/maven_jar 目录中了。把 JAR 包下载到本地的好处就是当编译时会优先从本地的 JAR 包中找，如果本地存在就直接拿来用；如果不存在，就从 Maven 的中心仓库下载。

第一次执行 mvn compile 和 mvn clean 这两个命令时，Maven 会去中心仓库下载需要的 JAR 包；而第二次执行这两个命令时，由于所需的 JAR 包已经在本地的仓库中存储，因此就可以直接使用了，这样就省了去中心仓库下载 JAR 包。

11.2.5 Maven 的简单使用

Maven 项目的目录约定如下。
```
Maven Project Root(项目根目录)
     |----src
     |    |----main
     |    |    |----java         存放项目的.java 文件
     |    |    |----resources    存放项目资源文件，如 spring、hibernate 配置文件
     |    |----test
     |    |    |----java         存放所有测试.java 文件，如 JUnit 测试类
     |    |    |----resources    存放项目资源文件，如 spring、hibernate 配置文件
     |----target                 项目输出位置
     |----pom.xml                用于标识该项目是一个 Maven 项目
```

11.2.6 pom.xml 文件中的 groupId 和 artifactId 到底该怎么定义

可以从 mvnrepository 官网中查找开源模块的依赖配置，找到合适的版本，然后复制其中的 Maven 配置到 Maven 项目的 pom.xml 文件里的<dependency></dependency>节点中，如图 11-16 所示。

图 11-16 开源模块的依赖配置

11.2.7 常用 Maven 命令

常用 Maven 命令及其描述如表 11-1 所示。

表 11-1　　　　　　　　　　　常用 Maven 命令及其描述

常用 Maven 命令	描述
mvn compile	编译源代码
mvn test	运行测试
mvn package	打包
mvn clean	清除产生的项目

11.3 配置 IntelliJ IDEA

IntelliJ IDEA 是 Java 编程语言开发的集成环境。IntelliJ IDEA 是一种商业化销售的 Java 集成开发环境，由 JetBrains 软件公司开发，提供 Apache 2.0 开放式授权的社区版本以及专有软件的商业版本，开发者可根据其所需选择下载使用。

本节介绍 IntelliJ IDEA 的常用配置和创建项目。

11.3.1 配置 JDK

启动 IntelliJ IDEA，选择"File"→"Project Structure"，弹出"Project Structure"对话框，单击左侧标签页，单击"SDKs"，如图 11-17 所示。

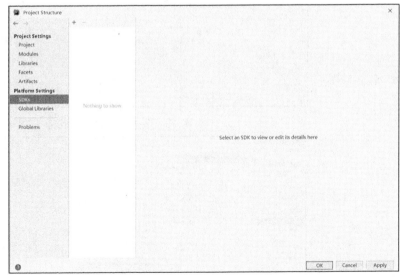

图 11-17　配置 SDKs

单击"+"，选择"JDK"，如图 11-18 所示。
在弹出的对话框中选择 JDK 的安装路径，如图 11-19 所示。
单击"OK"按钮，就配置完成 JDK 了，如图 11-20 所示。

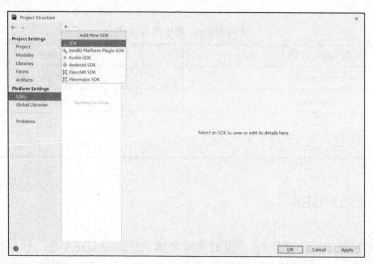

图 11-18　添加 JDK

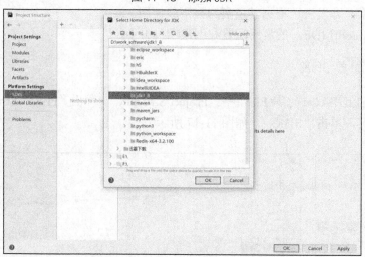

图 11-19　选择 JDK 的安装路径

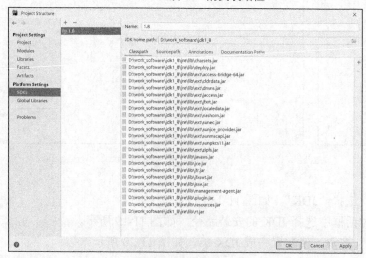

图 11-20　配置完成 JDK

11.3.2 配置 Maven

启动 IntelliJ IDEA，单击左上角的"File"，在下拉菜单中选择"Settings"选项，如图 11-21 所示。

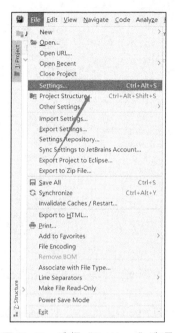

图 11-21 选择"Settings"选项

在弹出的"Settings"对话框中，在搜索文本框里输入 maven，然后选择"Maven"选项，如图 11-22 所示。

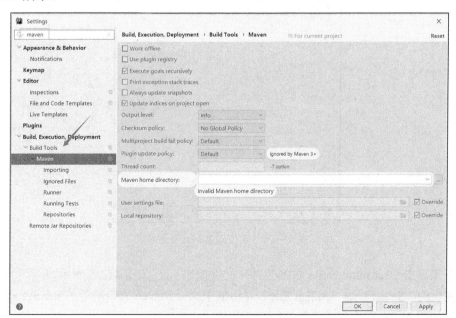

图 11-22 选择"Maven"选项

接下来单击右侧的"Maven home directory",如图 11-23 所示。

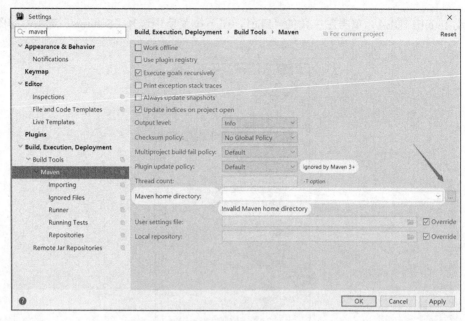

图 11-23　单击"Maven home directory"

然后在弹出的"Select Maven Home Directory"对话框中定位到 Maven 安装包,如图 11-24 所示。

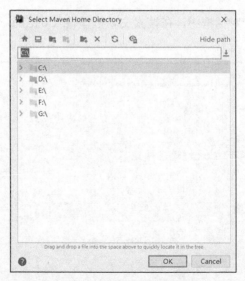

图 11-24　定位到 Maven 安装包

接下来单击 User settings file 右侧的打开文件按钮 ，如图 11-25 所示。

在弹出的"Select User Settings File"对话框中选择 Maven 安装包下的 settings.xml 配置文件即可,如图 11-26 所示。

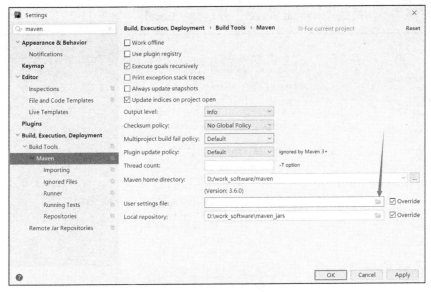

图 11-25 单击打开文件

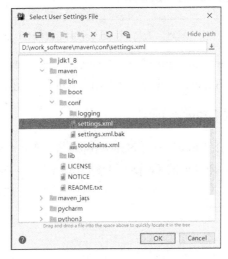

图 11-26 选择 Maven 安装包下的 settings.xml 配置文件

11.3.3 配置 Tomcat

选择"Run"→"Edit Configurations",如图 11-27 所示。

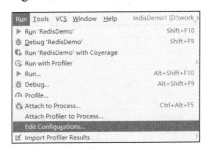

图 11-27 选择"Edit Configurations"选项

单击左侧"+",找到"Tomcat Server"并选择"Local",如图 11-28 所示。

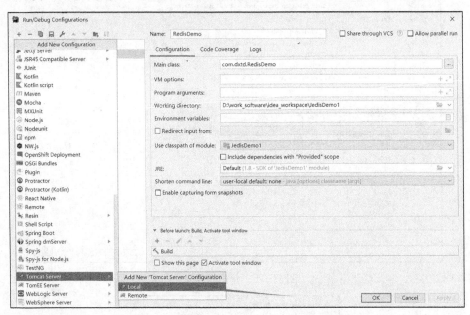

图 11-28　找到"Tomcat Server"并选择"Local"

在"Tomcat Server"→"Unnamed"→"Server"→"Application Server"下,单击"Configuration",找到本地 Tomcat 服务器,再单击"OK"按钮。

11.3.4　创建简单的 Maven 项目

启动 IntelliJ IDEA,选择"File"→"New Project",弹出"New Project"对话框,单击左侧标签页的"Maven",选择合适的 Project SDK,如图 11-29 所示。

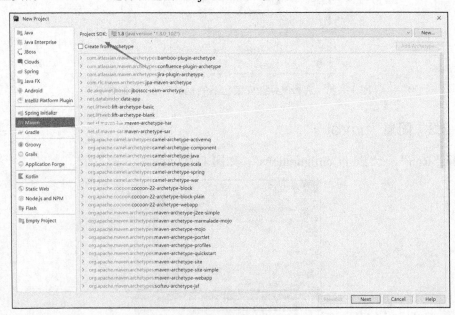

图 11-29　选择合适的 Project SDK

单击"Next"按钮，继续下一步的操作。

设置 GroupId 和 ArtifactId。GroupId 代表公司名称，ArtifactId 代表项目名称。Maven 的坐标为各种构件引入了秩序，任何一个构件都必须明确定义自己的坐标，Maven 的坐标包括如下的元素。

- GroupId：定义当前 Maven 项目隶属的实际项目。
- ArtifactId：该元素定义实际项目中的一个 Maven 项目或模块。

在本例中设置 GroupId 为 com.dxtd，ArtifactId 为 TestMaven。然后单击"Next"按钮，继续下一步的操作，如图 11-30 所示。

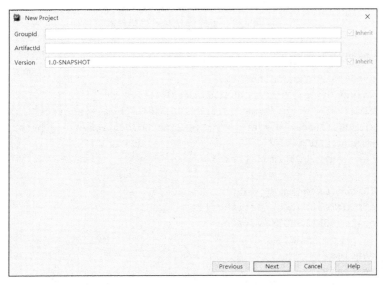

图 11-30　设置 GroupId 和 ArtifactId

选择合适的 Project location（项目位置），如图 11-31 所示。然后单击"Finish"按钮，生成 Maven 项目。生成的 Maven 项目结构如图 11-32 所示。

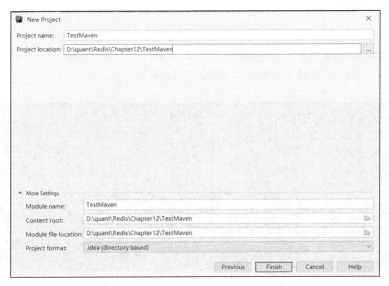

图 11-31　设置 Project name 和 Project location

图 11-32　生成的 Maven 项目结构

pom.xml 文件中的内容如下。

```xml
<?xml version="1.0" encoding="UTF-8"?>
<project xmlns="http://maven.apache.org/POM/4.0.0"
         xmlns:xsi="http://www.w3.org/2001/XMLSchema-instance"
         xsi:schemaLocation="http://maven.apache.org/POM/4.0.0 http://maven.apache.org/xsd/maven-4.0.0.xsd">
    <modelVersion>4.0.0</modelVersion>

    <groupId>com.dxtd</groupId>
    <artifactId>TestMaven</artifactId>
    <version>1.0-SNAPSHOT</version>

    <dependencies>
        <dependency>
            <groupId>junit</groupId>
            <artifactId>junit</artifactId>
            <version>3.8.1</version>
        </dependency>
    </dependencies>

</project>
```

使用 Maven 命令编译整个 Maven 项目。进入项目根目录，然后使用 Maven 命令"mvn compile"对整个 Maven 项目进行编译，如图 11-33 所示。

```
D:\quant\Redis\Chapter12\TestMaven>mvn compile
[INFO] Scanning for projects...
[INFO]
[INFO] ------------------< com.dxtd:TestMaven >------------------
[INFO] Building TestMaven 1.0-SNAPSHOT
[INFO] --------------------------[ jar ]--------------------------
[INFO]
[INFO] --- maven-resources-plugin:2.6:resources (default-resources) @ TestMaven ---
[WARNING] Using platform encoding (GBK actually) to copy filtered resources, i.e. build is platform dependent!
[INFO] Copying 0 resource
[INFO]
[INFO] --- maven-compiler-plugin:3.1:compile (default-compile) @ TestMaven ---
[INFO] Nothing to compile - all classes are up to date
[INFO]
[INFO] BUILD SUCCESS
[INFO]
[INFO] Total time:  1.383 s
[INFO] Finished at: 2019-10-05T01:15:27+08:00
[INFO]
```

图 11-33　使用 Maven 命令编译项目

使用 Maven 的"mvn compile"命令编译项目完成之后，在项目根目录下会生成一个 target 文件夹，如图 11-34 所示。

打开 target 文件夹，可以看到里面有一个 classes 文件夹，classes 文件夹存放的就是编译成功后生成的扩展名为.class 的文件。

使用 mvn clean 命令清除编译结果，也就是把编译生成的 target 文件夹删除。

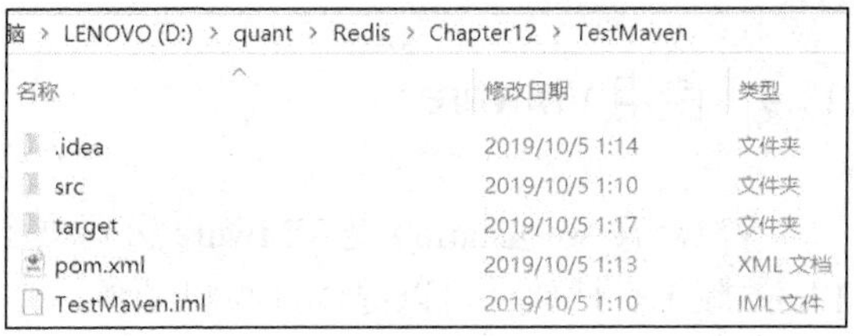

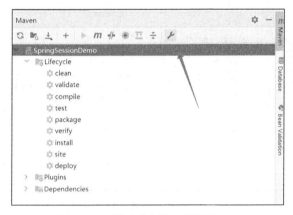

图 11-34　编译后的目录结构

11.3.5　导入 Maven 项目进行配置

使用 IntelliJ IDEA 导入 Maven 项目后，单击右侧的工具栏 Maven，如图 11-35 所示。

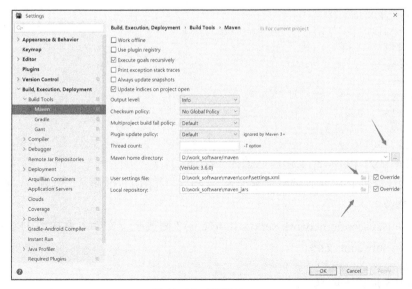

图 11-35　单击右侧的工具栏 Maven

然后在弹出的"Settings"对话框中，配置 Maven home directory、User settings file 和 Local repository，如图 11-36 所示。

图 11-36　配置 Maven

11.4 使用 VMware

VMware Workstation 是 VMware 公司推出的一款桌面虚拟计算机软件，具有 Windows、Linux 版本。此软件可以提供虚拟机功能，使计算机同时运行多个不同的操作系统。

11.4.1 配置虚拟机的静态 IP 地址

配置虚拟机的静态 IP 地址需要设置网络连接方式为 NAT 模式，如图 11-37 所示。

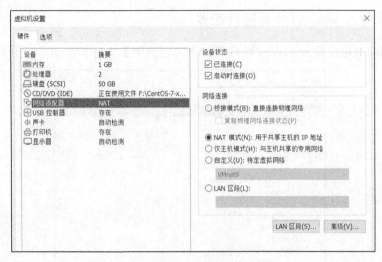

图 11-37 设置虚拟机的网络连接方式

编辑文件 ifcfg-ens33。
```
vi /etc/sysconfig/network-scripts/ifcfg-ens33
```
修改如下内容。
```
TYPE=Ethernet
BOOTPROTO=static
NAME=ens33
DEVICE=ens33
ONBOOT=yes
IPADDR=192.168.11.10
GATEWAY=192.168.11.2
NETMASK=255.255.255.0
DNS1=192.168.11.2
```
设置 IP 地址为 192.168.11.10，网关为 192.168.11.2。

在笔者的计算机上 IP 地址的网关地址为 56 网段，如图 11-38 所示，读者可以根据自己的实际情况进行修改。

使用 vi /etc/sysconfig/network-scripts/ifcfg-ens33 修改配置文件，添加图 11-39 所示内容。

修改/etc/resolv.conf 文件。
```
vi /etc/resolv.conf
```
在文件中添加以下内容。
```
nameserver 192.168.56.2
```

第 11 章 扩展知识

然后保存文件，最后重启网络。

```
service network restart
```

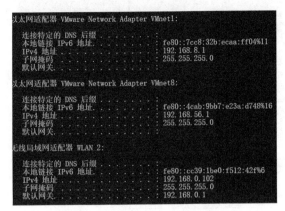

图 11-38　查看 VMnet8 的 IP 地址网段

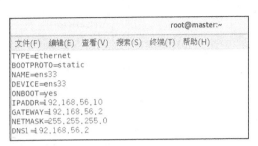

图 11-39　配置虚拟机的静态 IP 地址

11.4.2　恢复网络设置

如果 VMware 虚拟机出现问题，可以尝试恢复网络设置。打开 VMware，选择"编辑"→"虚拟网络编辑器"，然后单击"还原默认设置"按钮恢复虚拟机的网络设置，如图 11-40 所示。

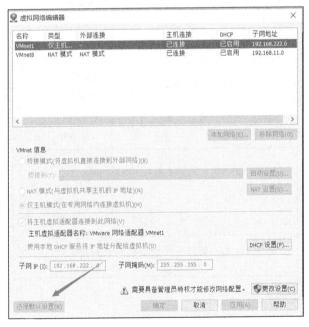

图 11-40　恢复虚拟机的网络设置

11.4.3　重新生成虚拟机网卡的 MAC 地址

为了保证每个 Linux 虚拟机的独立性和完整性，每个 Linux 虚拟机都需要一个独立的 MAC 地址，在 VMware 创建的 Linux 虚拟机中单击"编辑虚拟机设置"，如图 11-41 所示。

255

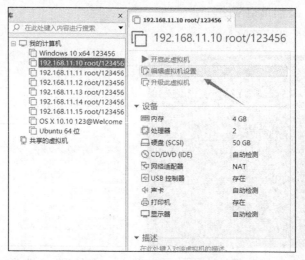

图 11-41　编辑虚拟机设置

在"虚拟机设置"对话框中单击"高级"按钮,在弹出的"网络适配器高级设置"对话框中,在"MAC 地址"下单击"生成"按钮,重新生成虚拟机网卡的 MAC 地址,如图 11-42 所示。

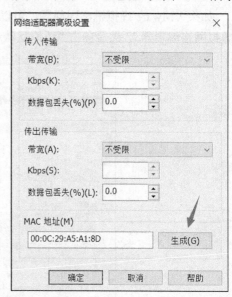

图 11-42　重新生成虚拟机网卡的 MAC 地址

11.5　配置 SecureCRT

SecureCRT 是一款支持 SSH(SSH1 和 SSH2)的终端仿真程序,同时支持 Telnet 和 rlogin 协议。SecureCRT 是一款用于连接运行包括 Windows、UNIX 和 VMS 的远程操作系统的理想工具。它可以使用内含的 VCP 命令行程序进行加密文件的传输。SecureCRT 有流行 CRT Telnet 客户端的所有特点,能从命令行中运行或从浏览器中运行。其他特点包括文本手稿、易于使用的工具条、用户的键位图编辑器、可定制的 ANSI 颜色等。SecureCRT 的 SSH 协议支持 DES、

3DES 和 RC4 密码与 RSA 鉴别。

11.5.1 设置打开的连接显示在一个页面

单击 SecureCRT 的"连接"按钮，在"连接"窗口勾选"在标签页中打开"，该选项可以使 SecureCRT 打开的多个连接显示在一个页面，如图 11-43 和图 11-44 所示。

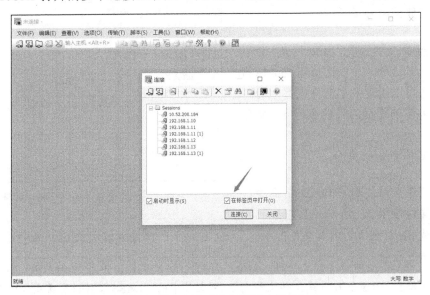

图 11-43 勾选"在标签页中打开"

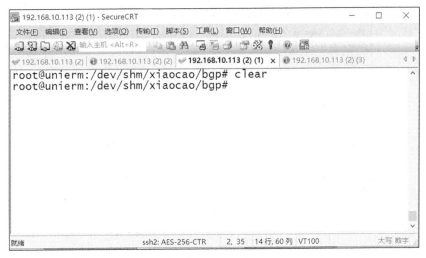

图 11-44 SecureCRT 打开的多个连接显示在一个页面

11.5.2 传输文件和下载文件

使用 SecureCRT 连接 CentOS 时需要安装 lrzsz 模块。

```
$ yum install lrzsz
```

设置上传路径和下载路径。打开 SecureCRT。选择"选项"→"会话选项"，弹出"会话选项"对话框，选择终端"X/Y/Zmodem"，如图 11-45 所示。

上传路径：使用 rz 命令进行上传文件操作时，弹出的对话框会默认定位到该目录下。

下载路径：使用 sz 命令进行下载文件操作后，下载的文件会默认下载到该目录下。

上传文件使用 rz 命令。

```
$ rz
```

输入 rz 命令然后按"Enter"键，会出现文件选择对话框，可以选择需要上传的文件，一次可以指定多个文件进行上传。上传到服务器的路径为当前执行 rz 命令的目录。

下载文件使用 sz 命令。

```
$ sz {filename}
```

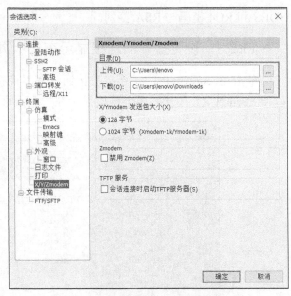

图 11-45　设置上传路径和下载路径

下载一个文件使用 sz filename 命令，下载多个文件使用 sz filename1 filename2 命令，下载 dir 目录下的所有文件但不包含 dir 目录下的文件夹使用 sz dir/*命令。

使用 rz 命令上传文件时，如果上传的是 JAR 文件，并且 SecureCRT 客户端上传时勾选了"以 ASCII 方式上传文件"，JAR 文件中的 CLASS 文件就会丢失。所以上传 JAR 文件时不能勾选"以 ASCII 方式上传文件"，如图 11-46 所示。

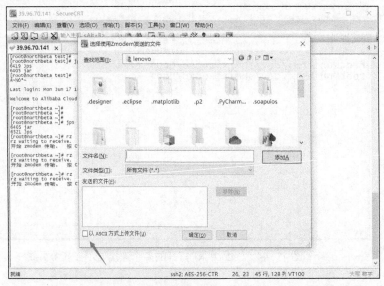

图 11-46　上传 JAR 文件时不能勾选"以 ASCII 方式上传文件"

11.5.3　显示中文

打开 SecureCRT，选择"选项"→"会话选项"，选择左侧树状列表中的"终端"→"外观",选择字符编码为 UTF-8,即可显示中文,如图 11-47 所示。

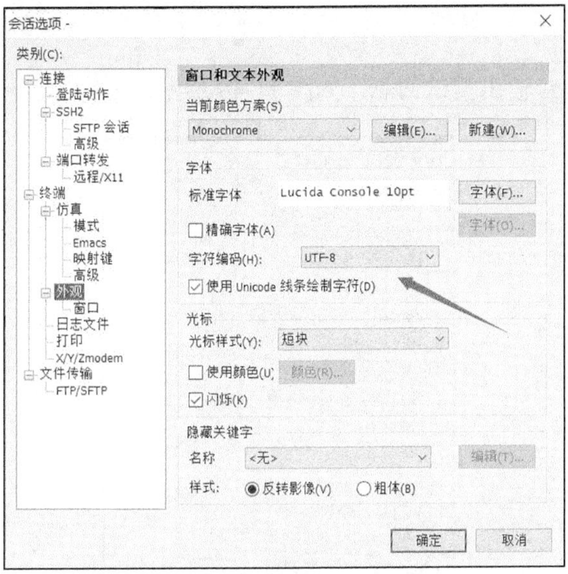

图 11-47 设置 SecureCRT 的字符编码

11.6 Chrome 的常用技巧

Chrome 是由 Google 公司开发的免费网页浏览器。Chrome 的特点是简洁、快速。此外,Chrome 基于更强大的 JavaScript V8 引擎,这是当前 Web 浏览器所无法实现的。

本节主要介绍 Chrome 的常用技巧。

11.6.1 打开开发者工具控制台

打开 Chrome,按"F12"键打开控制台,或按"Ctrl+Shift+J"组合键切换到 Console 窗口。

11.6.2 基本输出

用 console.log()方法在控制台输出内容。

```
//在控制台输出自定义字符串
console.log("输出字符串");

//在控制台输出自定义错误信息
console.error("我是一个错误");

//在控制台输出自定义信息
console.info("我是一条信息");

//在控制台输出自定义警告信息
console.warn("我是一个警告");

//在控制台输出自定义调试信息
```

```
console.debug("我是一个调试");

//清空控制台
console.clear();
```
在 Chrome 的控制台输出测试信息，如图 11-48 所示。

图 11-48　在 Chrome 的控制台输出测试信息

11.6.3　Chrome 禁用缓存

在 Chrome 中按"F12"键或按"Ctrl + Shift + J"或"Ctrl + Shift + I"组合键，选择"Network"选项，勾选"Disable cache"，如图 11-49 所示。

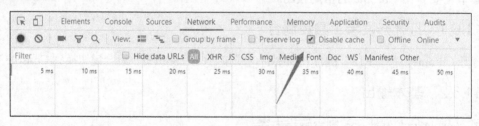

图 11-49　Chrome 禁用缓存

11.7　使用 Python 3 操作 Redis 集群

在 7.4 节中主要介绍了使用 Java 操作 Redis 集群，作为内容的补充，本节使用 Python 3 来操作 Redis 集群。

11.7.1　在 Windows 下安装 Python 3

本小节讲解在 Windows 下安装并配置 Python 3 开发环境。Python 安装环境信息如表 11-2 所示。

表 11-2　　　　　　　　　　　　Python 安装环境信息

操作系统	Windows 10　64 位
Python	3.6.4

访问 Python 官网，在下载页面中下载所需要的 Python 3.6.4 的安装包，读者可根据自己使用的操作系统选择相应的版本进行下载，如图 11-50 所示。对 Windows 用户来说，如果是 32 位操作系统，则选择 x86 版本；如果是 64 位操作系统，则选择 x86-64 版本。下载完成后，会得到一个以.exe 为扩展名的文件，双击该文件进行安装。

图 11-50　下载 Python 3.6.4

选择自定义安装 Python 3.6.4，如图 11-51 和图 11-52 所示。安装路径可以自己决定，笔者的安装路径是 D:\installed_software\python3，如图 11-53 所示。

　　在安装过程中按照提示一步步操作就行，但安装路径尽量不要带有中文或空格，以避免在使用过程中出现一些莫名的错误。

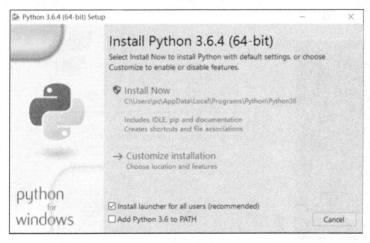

图 11-51　选择"Customize installation"自定义安装

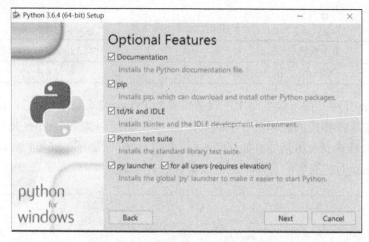

图 11-52　安装 Python 3.6.4 的可选项

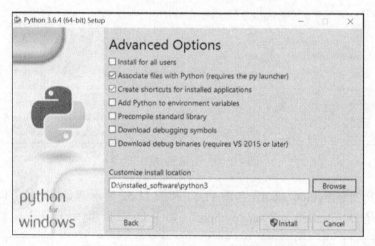

图 11-53　设置 Python 3.6.4 的安装路径

安装完成后,可以在"开始"菜单中看到 Python 3.6 目录,如图 11-54 所示。

打开 Python 自带的 IDLE (Python 3.6 64-bit),就可以编写 Python 程序了。Python 3.6.4 Shell 页面如图 11-55 所示。

图 11-54　Python 3.6 目录

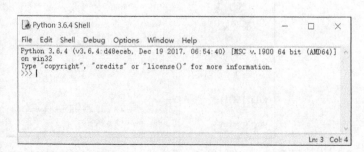

图 11-55　Python 3.6.4 Shell 页面

IDLE 是一个 Python Shell，Shell 的意思是"外壳"，就是一个通过键入文本与程序交互的途径。

还需要把 Python 的安装目录添加到系统变量 Path 中，在桌面上右击"此电脑"，弹出快捷菜单，选择"属性"→"高级系统设置"→"高级"，单击"环境变量"按钮，弹出"环境变量"对话框，如图 11-56 所示。

图 11-56　设置系统变量 Path

在系统变量 Path 中添加变量值。
`d:/installed_software/python3;d:/installed_software/python3/Scripts;`

（1）d:/installed_software/Python3 是笔者在本机上安装 Python 3 的位置，读者需要根据自己计算机上的实际情况进行修改。
（2）在 Windows 下路径中的变量之间使用分号分隔。

添加变量值成功后，如图 11-57 所示。

还需要配置系统变量 PYTHONPATH，用来永久设置模块的搜索路径。使用 pip 命令安装第三方模块时，第三方模块存放在%/python3/Lib/site-packages 目录下，所以需要把%/python3/Lib/site-packages 对应的目录添加到系统变量 PYTHONPATH 中，如图 11-58 所示。

在桌面上右击"此电脑"，弹出快捷菜单，选择"属性"→"高级系统设置"→"高级"，单击"环境变量"按钮，弹出"环境变量"对话框。在系统变量 PYTHONPATH 中添加变量值。
`d:/install_software/python3/Lib/site-packages;`

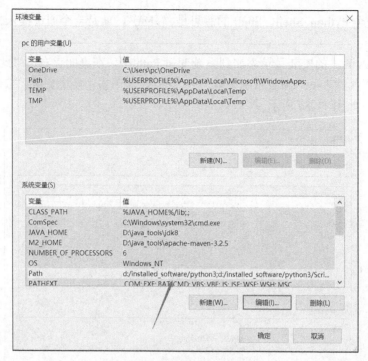

图 11-57　配置系统变量 Path

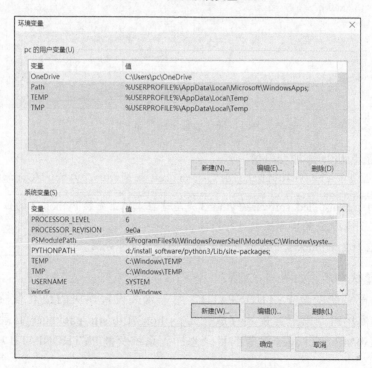

图 11-58　配置系统变量 PYTHONPATH

 d:/install_software/python3/Lib/site-packages 是在笔者的计算机上安装 Python 3 的第三方模块的配置路径，读者需要根据自己计算机上的实际情况进行修改。

现在，我们检验一下 Python 3 是否安装成功。按"Win+R"组合键运行 cmd 命令，进入命令提示符窗口，如图 11-59 所示。

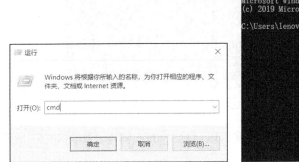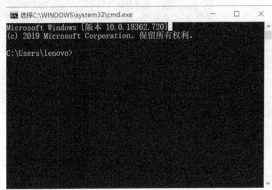

图 11-59　进入命令提示符窗口

在命令行输入 python，开始启动 Python IDLE，需要几秒的时间。启动后，读者就可以看到它的页面中包含一个交互式终端，也可以看到所安装的 Python 版本号，如图 11-60 所示。这时 Python 3 的运行环境就安装好了。

图 11-60　Python 3 的运行环境

Python 的行首显示的 3 个大于号（>>>）是命令提示符，当看到这个命令提示符时，就表示解释器正在等待输入命令。下面尝试在命令提示符后输入以下命令。

```
print("hello python")
```

按"Enter"键，Python 就会执行所输入的命令，并在窗口中显示运行结果。也可以把命令行看成计算器来计算表达式的值，把在命令提示符后面输入的每一条命令看成一个程序，命令行每次只运行这个程序中的一行；还可以在命令行中创建变量或导入模块，如图 11-61 所示。

图 11-61　创建变量、计算和导入模块

import 命令把 Python 数学函数库的功能导入程序中。上面的程序使用了变量和赋值运算符"="，其含义是对 9 开平方，并把结果赋值给 r，最后把结果输到屏幕上。

可以通过 help 命令获取某个函数的使用方法，如图 11-62 所示。

```
help(print)
```

如果想退出命令行模式，按"Ctrl + C"组合键。

还可以把 Python 代码写在一个扩展名为.py 的文件里，这个文件就叫脚本文件。生成一个名为 hello.py 的文件，包含以下内容。

```
print("Hello World!")
```

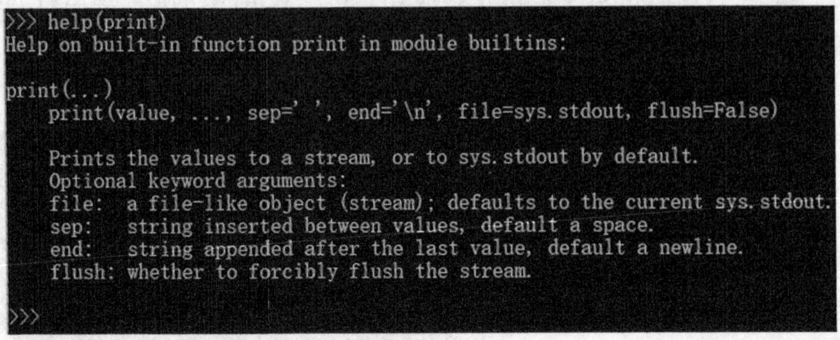

```
>>> help(print)
Help on built-in function print in module builtins:

print(...)
    print(value, ..., sep=' ', end='\n', file=sys.stdout, flush=False)
    
    Prints the values to a stream, or to sys.stdout by default.
    Optional keyword arguments:
    file:  a file-like object (stream); defaults to the current sys.stdout.
    sep:   string inserted between values, default a space.
    end:   string appended after the last value, default a newline.
    flush: whether to forcibly flush the stream.
>>>
```

图 11-62　通过 help 命令获取某个函数的使用方法

进入 hello.py 文件，然后输入以下命令行，就可以运行 Python 脚本，如图 11-63 所示。

```
python hello.py
```

```
G:\quant2\Python\Chapter01>python hello.py
Hello World!
```

图 11-63　运行 Python 脚本

11.7.2　在 Linux 下安装 Python 3

本小节讲解在 Linux 下安装并配置 Python 3 开发环境。本书使用的 Linux 平台是 CentOS 7，安装环境信息如表 11-3 所示。

表 11-3　　　　　　　　　　　Python 3 安装环境信息

操作系统	CentOS 7　64 位
Python	3.6.4

1．安装 Python 3 前的库环境准备

本小节从源代码安装 Python 3，从源代码编译和安装 Python 需要 C/C++ 编译器和其他开发包。因此，安装 Python 3 前首先需要安装相关开发包，准备好库环境。可以使用如下命令，准备安装 Python 3 前的库环境。

```
$ yum install gcc patch libffi-devel python-devel  zlib-devel bzip2-devel openssl-devel ncurses-devel sqlite-devel readline-devel tk-devel gdbm-devel db4-devel libpcap-devel xz-devel -y
```

2．从源码包安装 Python 3

在 Python 官网下载 Python 3 的源码包，本书使用的版本是 Python 3.6.4，对应的文件是 Python-3.6.4.tgz。

首先切换到 root 用户。root 用户是 Linux 系统的最高权限用户，因为 root 用户权限过高，系统一般情况下不允许使用 root 用户登录系统。但是以普通用户登录系统后，普通用户权力受限，无法完成一些对系统有重大影响的操作，比如安装应用程序。所以这里需要切换到 root 用户来安装 Python 3。

使用如下命令从普通用户切换到 root 用户。
```
$ su root
```
按 Enter 键后，系统提示输入 root 密码。输入密码验证通过后，切换 root 用户完成。
然后，使用如下命令建立安装目录/software。
```
$ mkdir /software
```
把 Python-3.6.4.tgz 文件上传到/software 目录下，并解压缩文件。
```
$ cd /software
$ tar -zxvf Python-3.6.4.tgz
```
进入解压缩后的文件夹。
```
$ cd Python-3.6.4/
```
Python 3 的默认安装目录是/usr/local，如果要改成其他目录可以在编译前使用 configure 命令并添加参数"-prefix=/usr/local/python"来完成修改，本小节指定 Python 3 的安装目录为/usr/local/python。
```
$ ./configure -prefix=/usr/local/python
```
接着，编译 Python 源码。
```
$ make
```
最后，执行安装。
```
$ make install
```
至此已经在 CentOS 7 操作系统中成功安装了 Python 3。但还需要把 Python 3 的配置信息添加到 Linux 的环境变量 PATH 中。编辑/etc/profile 文件。
```
$ vi /etc/profile
```
添加以下内容到文件末尾，然后保存文件，退出到命令行。
```
export PATH=/usr/local/python/bin:..:$PATH
```
最后，激活/etc/profile 文件。
```
$ source /etc/profile
```
在命令行输入命令 python3 就可以调用 Python 3 了，如图 11-64 所示。

```
[root@bogon ~]# python3
Python 3.6.4 (default, Jun 21 2018, 03:10:48)
[GCC 4.8.5 20150623 (Red Hat 4.8.5-28)] on linux
Type "help", "copyright", "credits" or "license" for more information.
>>>
```

图 11-64　输入命令 python3 调用 Python 3

　　/usr/local 下一般是安装软件的目录，笔者建议把 Python 3.6.4 安装到/usr/local/python 目录下。
　　Linux 系统变量的路径之间使用冒号（:）相连。

11.7.3　使用 Redis 模块

Python 标准模块中没有操作 Redis 的模块，但已经有开源的操作 Redis 单机的 Python 模块：redis-py 和连接 Redis 集群的 Python 模块：redis-py-cluster。

安装 Python 操作 Redis 单机模块，需要使用以下 pip 命令。
```
pip install redis
```
安装 Python 操作 Redis 集群模块，需要使用以下 pip 命令。

```
pip install redis-py-cluster
```
命令结果如图 11-65 所示。

```
C:\Users\lenovo> pip install redis-py-cluster
Collecting redis-py-cluster
  Downloading https://files.pythonhosted.org/packages/35/cb/29d44f7735af4fe9251afb6b5b173ec79c6e8f49cb6a61603e77a54ba658
/redis_py_cluster-2.0.0-py2.py3-none-any.whl
Collecting redis<3.1.0,>=3.0.0 (from redis-py-cluster)
  Downloading https://files.pythonhosted.org/packages/f5/00/5253aff5e747faf10d8ceb35fb5569b848cde2fdc13685d42fcf63118bbc
/redis-3.0.1-py2.py3-none-any.whl (61kB)
     |████████████████████████████████| 71kB 4.0kB/s
Installing collected packages: redis, redis-py-cluster
  Found existing installation: redis 3.2.1
    Uninstalling redis-3.2.1:
      Successfully uninstalled redis-3.2.1
Successfully installed redis-3.0.1 redis-py-cluster-2.0.0
```

图 11-65 安装 Python 操作 Redis 集群模块的命令

请确认 Redis 服务器已经启动并运行,具体内容请参考 1.2 节 Redis 环境搭建。新建 Python 脚本 RedisDemo.py,操作 Redis 单机。

本实例文件名为 Redis\Chapter11\RedisDemo.py,内容如下。

```python
import redis

r = redis.Redis(host='localhost', port=6379, db=0)
r.set('password', 'abcdef')

print(r.get('password'))
```

程序会在 Redis 的数据库 0 中存储一个 key-value 键值对。运行程序得到如下结果。

```
b'abcdef'
```

请确认 Redis 集群已经启动并运行,具体内容请参考 6.3 节 Redis 集群。新建 Python 脚本 RedisCluster.py,操作 Redis 集群。

本实例文件名为 Redis\Chapter11\RedisCluster.py,内容如下。

```python
from rediscluster import RedisCluster
# Redis 集群各节点的主机 IP 地址和端口
cluster_nodes = [{'host': '192.168.11.15', 'port': 8001},
                 {'host': '192.168.11.15', 'port': 8002},
                 {'host': '192.168.11.15', 'port': 8003},
                 {'host': '192.168.11.15', 'port': 8004},
                 {'host': '192.168.11.15', 'port': 8005},
                 {'host': '192.168.11.15', 'port': 8006}]

cluster = RedisCluster(startup_nodes=cluster_nodes)
cluster.set('book', 'Python')
print(cluster.get('password'))
```

运行程序得到如下结果。

```
b'Python'
```

使用 redis-cli 命令连接主机 192.168.11.15 上的 Redis 集群节点。

```
[root@localhost ~]# redis-cli -c -h 192.168.11.15 -p 8001
192.168.11.15:8001> GET book
"Python"
```

在 Redis 客户端查询 book 获得的结果和程序运行结果一致。